Springer Theses

Recognizing Outstanding Ph.D. Research

Aims and Scope

The series "Springer Theses" brings together a selection of the very best Ph.D. theses from around the world and across the physical sciences. Nominated and endorsed by two recognized specialists, each published volume has been selected for its scientific excellence and the high impact of its contents for the pertinent field of research. For greater accessibility to non-specialists, the published versions include an extended introduction, as well as a foreword by the student's supervisor explaining the special relevance of the work for the field. As a whole, the series will provide a valuable resource both for newcomers to the research fields described, and for other scientists seeking detailed background information on special questions. Finally, it provides an accredited documentation of the valuable contributions made by today's younger generation of scientists.

Theses are accepted into the series by invited nomination only and must fulfill all of the following criteria

- They must be written in good English.
- The topic should fall within the confines of Chemistry, Physics, Earth Sciences, Engineering and related interdisciplinary fields such as Materials, Nanoscience, Chemical Engineering, Complex Systems and Biophysics.
- The work reported in the thesis must represent a significant scientific advance.
- If the thesis includes previously published material, permission to reproduce this must be gained from the respective copyright holder.
- They must have been examined and passed during the 12 months prior to nomination.
- Each thesis should include a foreword by the supervisor outlining the significance of its content.
- The theses should have a clearly defined structure including an introduction accessible to scientists not expert in that particular field.

Indexed by zbMATH.

More information about this series at http://www.springer.com/series/8790

Eileen Otte

Structured Singular Light Fields

Doctoral Thesis accepted by
University of Münster, Münster, Germany

Springer

Author
Dr. Eileen Otte
Institute of Applied Physics
University of Münster
Münster, Germany

Supervisors
Prof. Cornelia Denz
Institute of Applied Physics
University of Münster
Münster, Germany

Prof. Andrew Forbes
School of Physics
University of the Witwatersrand
Johannesburg, South Africa

ISSN 2190-5053 ISSN 2190-5061 (electronic)
Springer Theses
ISBN 978-3-030-63714-9 ISBN 978-3-030-63715-6 (eBook)
https://doi.org/10.1007/978-3-030-63715-6

This Springer imprint is published by the registered company Springer Nature Switzerland AG
The registered company address is: Gewerbestrasse 11, 6330 Cham, Switzerland

Supervisors' Foreword

Structured singular light fields represent a ubiquitous phenomenon in nature which can not only be observed in shallow water but also in the sunlight of the blue sky or in rainbows. These light fields are characterized by a spatially varying amplitude, phase, or polarization, being accompanied by the occurrence of singularities.

Optical singularities are particularly interesting since they are structurally stable units of the light field and they determine its topology. In ray optics, such singularities represent caustics of light, well known for centuries. In wave optics, optical singularities occur in the phase or polarization of light—which are of main interest in this book. Singularities in phase correspond to locations, in which the phase of light is undefined and intensity vanishes. Here, an optical vortex is observed, in which—in analogy to water vortices—the phase of light orbits around a dark central area. Polarization singularities are much less known, however, show an incredibly rich variety of different types of singularities. In these cases, the geometrical properties of the corresponding polarization ellipses cannot be defined, for instance, the direction of electric field oscillation, which leads to singularity points, lines, and curves.

These singularities form the invariant fundamental structure of the light field—its skeleton—which stays stable even for variations in the surrounding "tissue," under deformation or disturbances. This highlights the fundamental significance of optical singularities as a basis for the description of topical tailored light fields.

Even though structured singular light occurs naturally, its description and artificial realization represent a major current challenge. It was not until the late 1960s that singularities were systematically investigated in mathematics and theoretical physics, with the first experimental focus on the fine structures of caustics. The understanding of topological phase singularities has only started in the 1970s, revealing the remarkable properties of optical vortices. They give photons orbital angular momentum and propagate structurally stable even after disturbances. These properties made optical vortices or, more precisely, phase singularities attractive and are since only a few years known for applications in optical trapping or as a new spatial degree of freedom for optical information processing. Most famous was the application of an optical vortex-bearing beam to create nanoscopic superresolution.

An even younger field of research are polarization singularities which appear in structured vectorial light fields. These include, among others, lines of pure circular or surfaces of pure linear polarization in three-dimensional topologies, representing generic, most stable singularities. Experimentally, polarization was long seen as a traditional parameter with two states as, e.g., linear polarization. Thus, polarization singularities were up to now a field that was mostly considered theoretically, and systematic techniques for the customization of polarization structured light including its singularities were not present.

It is due to the thesis of Eileen Otte that polarization has been developed from a forgotten parameter of structured light to an important field of experimental singularity research and its applications. In particular, the interaction of phase-dependent orbital and polarization-dependent spin angular momentum plays a crucial role within these light fields, even enabling entanglement as well as spin-orbit coupling.

In this book, Eileen Otte comprehensively describes novel approaches enabling the full customization, detailed analysis, and application of structured singular light fields. She pioneers the characterization of topological singularities and even singularity networks, describing their underlying energy flow and their propagation behavior in singular fields. Furthermore, she presents tightly focused vectorial light fields that have striking three-dimensional polarization topologies. Among them, singularity networks and landscapes as polarization Möbius strip arrays impressively show the potential of polarization topology.

Eileen Otte also developed a pioneering nano-tomographic approach that allows analyzing these focused vectorial light fields and its typically invisible non-paraxial field properties by a single-shot method employing self-assembled molecular layers, thereby unlocking the immense potential of non-paraxial nano-structured light for applied optics.

Eileen Otte did not content herself with the advancement of understanding fundamental topological optics. She also paved the way to innovative applications: Structured singular light fields enable novel classical as well as quantum information technologies, and can be employed for tailored optical micro- and nano-manipulation.

The outstanding quality of Eileen's work is on the one hand proven by ten thesis-related publications in highly ranked international peer-reviewed journals but also by numerous awards. Her thesis was recognized by the excellent mark "summa cum laude," honored by the University of Muenster's Dissertation Award 2019, and the Research Award of the North Rhine-Westphalian Academy of Sciences, Humanities and the Arts funded by the Industrial Club Duesseldorf, Germany in 2020. She was also nominated for the young researcher's Dissertation Prize of the Physical Society of Germany in the field of Quantum Optics and Photonics.

Münster, Germany
Johannesburg, South Africa
August 2020

Cornelia Denz
Andrew Forbes

Publications Related to this Thesis

Scientific Publications

1. **E. Otte**, K. Tekce, S. Lamping, B. J. Ravoo, and C. Denz. Polarization nano-tomography of tightly focused topological light landscapes by self-assembled monolayers. *Nat. Commun.* 10.4308 (2019). [Sect. 4.2.1, 4.3].[1]
2. K. Tekce, **E. Otte**, and C. Denz. Optical singularities and Möbius strip arrays in non-paraxial light fields. *Opt. Express* 27 (2019), 29685. [Sect. 4.4].
3. **E. Otte**, E. Asché, and C. Denz. Shaping optical spin flow topologies by the translation of tailored orbital phase flow. *J. Opt.* 21 (2019), 064001. [Sect. 2.4.2].
4. **E. Otte** and C. Denz. Sculpting complex polarization singularity networks. *Opt. Lett.* 43 (2018), 5821. [Sect. 3.1].
5. **E. Otte**, I. Nape, C. Rosales-Guzmán, A. Vallés, C. Denz, and A. Forbes. Recovery of nonseparability in self-healing vector Bessel beams. *Phys. Rev. A* 98 (2018), 053818. [Sect. 3.3.1, 5.2].
6. **E. Otte**, K. Tekce, and C. Denz. Spatial multiplexing for tailored fully-structured light. *J. Opt.* 20 (2018), 105606. [Sect. 2.4.4].
7. **E. Otte**, C. Rosales-Guzmán, B. Ndagano, C. Denz, and A. Forbes. Entanglement beating in free space through spin-orbit coupling. *Light Sci. Appl.* 7 (2018), 18009. [Sect. 5.3].
8. **E. Otte**, C. Alpmann, and C. Denz. Polarization Singularity Explosions in Tailored Light Fields. *Laser Photonics Rev.* 12 (2018), 1700200. [Sect. 2.4.3, 3.2].
9. **E. Otte**, K. Tekce, and C. Denz. Tailored intensity landscapes by tight focusing of singular vector beams. *Opt. Express* 25 (2017), 20194. [Sect. 4.2.1].
10. **E. Otte**, C. Alpmann, and C. Denz. Higher-order polarization singularitites in tailored vector beams. *J. Opt.* 18 (2016), 074012. [Sect. 2.4.3].

Conference Proceedings

1. **E. Otte**, K. Tekce, and C. Denz. Customized focal light landscapes by complex vectorial fields for advanced optical trapping. *Proc. SPIE* 10549 (2018), 115. [Sect. 4.2.2, 4.2.3].

[1]In parts, results presented within this thesis have in parts been published in the listed publications. For involved publications, respective results can be found within the sections referred to in square brackets.

Journal Cover

1. **E. Otte**, C. Alpmann, and C. Denz. Polarization Singularity Explosions in Tailored Light Fields (Laser Photonics Rev. 12(6)/2018). *Laser Photonics Rev.* 12 (2018), 1870028—Inside front cover. [Sect. 2.4.3, 3.2].

Acknowledgements

The content of the thesis represents the results of several years of research in the *Nonlinear Photonics* group of Prof. Dr. Cornelia Denz as well as abroad studies in the *Structured Light Laboratory* of Prof. Dr. Andrew Forbes. To the success of respective research and, thereby, to the completion of this work various people contributed in different ways, for which I would like to express my sincerest gratitude.

In the first place, I thank Prof. Dr. Cornelia Denz for the opportunity to be part of her team and to work on the highly topical research field of structured singular light fields. The fantastic research environment, her scientific guidance, trust, and encouragement toward me have not only enabled the path from own ideas to scientific publications but have also contributed to my personal development. Her outstanding support allowed me to gain experience in team leading, student supervision, project coordination, meeting organization, proposal writing, and many more, from which I will significantly benefit in terms of my future career. In particular, I would like to thank her for diverse international experiences she has made possible, including the participation in several international, besides national, conferences and precious research stays abroad. Thank you for all these valuable experiences.

Second, I would like to thank Prof. Dr. Andrew Forbes not only for being my second supervisor but also especially for the precious and exciting possibility to be part of his group. The time I was allowed to spend in his group, several discussions and exciting projects, even ongoing cross-continental ones, have enriched my understanding of structured light, shedding new light on shaped beams, their quantum-like properties and applications. Thank you for your time, guidance, and inspiration. Further, I also want to thank his group members for always giving me the feeling of really belonging to the group. In particular, I thank Carmelo, Ben, Isaac, Bienvenu, Bernice, and Nkosi for fruitful discussions and the fun time, also outside the office and laboratory. I am looking forward to maintain international friendships.

My gratitude also goes to the various present and former members of the *Nonlinear Photonics* group for the helpful discussions, technical assistance, and nice atmosphere in meetings, coffee breaks, seminars, workshops, joint conferences, among others. Especially, I thank the students who worked with me on different aspects of this thesis, namely Kemal Tekce (Master), Ramon Runde (Master), and Eric Asché (Bachelor and student assistant). Thank you for your ideas, discussions, questions, and experimental assistance. Further, my thanks go to the "Tweezers Team"—I gratefully remember our intense beneficial weekly meetings, collaborations, and leisure activities.

For their deep and ongoing interest in my projects and critical discussions, I personally thank Christina, Kemal, Ramon, Álvaro, Eric, Carmelo, Isaac, Bienvenu, and Danica. Beyond, I want to thank Christina Alpmann for advising me at the beginning of my Ph.D. as well as for her infectious enthusiasm, and Álvaro Barroso Peña for technical assistance on zeolite-L trapping and fluorescence detection. I also thank the people who proofread this thesis.

Finally, I express my heartfelt gratitude to my family and Julian for their support, encouragement, and continuous motivation in every situation.

Cooperations

Partial content of this work originates from cooperation with students, colleagues, or scientific groups. These people are particularly acknowledged with their respective contributions listed below.

- Eric Asché is acknowledged for assistance on the experimental measurement of spin flow configurations presented in Sect. 2.4.2. Results in Fig. 2.10 have been obtained in cooperation with him [1]. [E. Asché. Complex energy flow structures in tailored light fields. Bachelor thesis. University of Muenster, 2016]
- Ramon Runde is acknowledged for his assistance on the study of vectorial discrete non-diffracting and self-imaging light fields in Sect. 3.3.2 and 3.3.3. Experimental results with respect to these fields have been obtained in cooperation with him. [R. Runde. Scalar and vectorial self-imaging light fields. Master thesis. University of Muenster, 2017]
- Kemal Tekce is acknowledged for his assistance on the realization of spatial multiplexing for fully-structuring light and the investigation of non-paraxial light fields in Sect. 2.4.4 and Chap. 4, respectively [2–6] . Respective experimental results have been obtained in cooperation with him. [K. Tekce. Three-dimensional light landscapes by tightly focusing phase and polarization structures. Master thesis. University of Muenster, 2017]
- Dr. Christina Alpmann is acknowledged for her assistance on investigations of 2d vector fields [7] (Sect. 2.4.3) and the preliminary experimental study of singularity explosions [8]
- Dr. Sebastian Lamping and Prof. Dr. Bart Jan Ravoo, Organic Chemistry Institute and Center for Soft Nanoscience, University of Muenster, are acknowledged for preparing and providing the self-assembled monolayer probes used within the experiments of Sect. 4.3 [5]

• Dr. Carmelo Rosales-Guzmán, Dr. Bienvenu Ndagano, and Isaac Nape of the group of Prof. Dr. Andrew Forbes, School of Physics, University of the Witwatersrand, Johannesburg, South Africa, are acknowledged for the assistance on the study of local entanglement/non-separability in structured light (Sect. 3.3.1, 5.2, 5.3). In particular, the supervision of these projects by Prof. Dr. Andrew Forbes is acknowledged [9, 10]

References

1. Otte E, Asché E, Denz C (2019) Shaping optical spin flow topologies by the translation of tailored orbital phase flow. J Opt 21:064001 (cit. on pp. g, 27, 30–34)
2. Otte E, Tekce K, Denz C (2017) Tailored intensity landscapes by tight focusing of singular vector beams. Opt Express 25:20194 (cit. on pp. g, 3, 83, 84, 89–91)
3. Otte E, Tekce K, Denz C (2018) Spatial multiplexing for tailored fully-structured light. J Opt 20: 105606 (cit. on pp. g, 4, 27, 38–43)
4. Otte E, Tekce K, Denz C (2018) Customized focal light landscapes by complex vectorial fields for advanced optical trapping. Proc SPIE 10549:115 (cit. on pp. g, 84, 89, 94)
5. Otte E, et al (2019) Polarization nano-tomography of tightly focused light landscapes by self-assembled monolayers. Nat Commun 10.4308 (cit. on pp. g, 84, 89, 92, 99–107)
6. Tekce K, Otte E, Denz C (2019) Optical singularities and Möbius strip arrays in tailored non-paraxial light fields. Opt Express 27:29685 (cit. on pp. g, 84, 108–115)
7. Otte E, Alpmann C, Denz C. (2016) Higher-order polarization singularitites in tailored vector beams. J Opt 18 : 074012 (cit. on pp. g, 2, 4, 23, 27, 30, 35–38)
8. Otte E, Alpmann C, Denz C (2018) Polarization Singularity Explosions in Tailored Light Fields. Laser Photonics Rev. 12:1700200 (cit. on pp. g, 15, 22, 24, 27, 30, 37, 38, 46, 56–59, 61–63)
9. Otte E, et al (2018) Entanglement beating in free space through spin-orbit coupling. Light Sci Appl 7:18009 (cit. on pp. g, 27, 28, 78, 118, 130–142)
10. Otte E, et al (2018) Recovery of nonseparability in self-healing vector Bessel beams. Phys Rev A 98:053818 (cit. on pp. g, 25, 46, 64, 66–68, 82, 118, 123–129)

Contents

Chapter 1
Introduction and Outline

> For any sceptic of the continued capacity of science to uncover new truth, to pave the way for previously unimagined applications, there is hardly a better corrective than [...] the science of light. It may be unscientific to say that light is unfathomable, but it certainly is a characteristic of the subject that there is always more to be learned, just when the utmost depths seem within grasp [1].

Light is an ubiquitous phenomenon. A glimpse of its various properties is already catched in the natural world: Rainbows and sunsets visualize light's broad colorful spectrum and its ability to refract. Further, photosynthesis, i.e., the process of converting the energy in sunlight to chemical energy used by plants, provides the proof of light being at the origin of all life. A crucial advancement in the science of light, essentially increasing its impact on human life, represents the invention of the laser—**L**ight **A**mplification by **S**timulated **E**mission of **R**adiation—by T. Maiman in 1960 [2]. Due to its valuable properties as its high energy, long coherence length, and tunable electromagnetic spectrum, laser light has become an indispensable tool in daily life. Industries as medical treatment and diagnostics, telecommunications, manufacturing, basic scientific research, and many more rely on lasers [3]. For instance, the feature to focus the high power of a laser to a pinpoint enables its use as a precision scalpel in medicine. For eye surgery or the removal of tumors, tattoos, hair or birthmarks the laser represents today's tool of choice. Furthermore, lasers have become an essential means for cutting through thick plates of steel or for faster internet speeds based on optical fiber technologies. Everyday, scientists and engineers continue to enhance known or develop novel laser-, i.e., light-based applications enriching our lives.

Contributing significantly to the advancement of laser technologies, *structured light* has come to the fore. This term as well as its equivalents shaped light, complex light, tailored light, sculpted light, among others, refer to the generation and application of customized light fields [4], being structured in amplitude, phase, and/ or polarization. Respective light fields have paved the way to various advanced appli-

E. Otte, *Structured Singular Light Fields*, Springer Theses,
https://doi.org/10.1007/978-3-030-63715-6_1

cations in optical manipulation [5–11], high-resolution imaging [12–14], material machining [15–17], metrology [18, 19], as well as classical and quantum communication [20–25] (see also Refs. [4, 26] and refs. therein). Arguably, one of the most famous examples might be Nobel prize awarded STED (stimulated emission depletion) microscopy [12]. Within this fluorescence based technique, the key to enhance spatial resolution represents the implementation of a beam revealing a transversely donut shaped intensity structure. More precisely, not the bright intensity itself determines transverse imaging resolution, but the on-axis embedded singular point of undefined phase—a phase singularity—and the corresponding dark center of the beam. The donut shaped beam depletes priorly excited fluorescence in its high intensity areas while leaving the subdiffractive central area active to emit fluorescence. Hence, not only structured light, but *structured singular light* facilitates this sophisticated high-resolution imaging method.

Singular light corresponds to the field of fundamental singular optics comprising the study of singularities in intensity, phase as well as polarization [27, 28]. The latter two are of main interest within this thesis. Interestingly, singular light is as ubiquitous as light itself: intensity singularities can be observed as bright lines at the bottom of shallow waters, phase singularities are hidden in optical speckle fields, and polarization singularities can be detected in the blue daylight sky [27, 29, 30]. While intensity singularities are found in the infinite brightness of caustics of light [29], scalar phase singularities, in contrast, which are observed in scalar light fields structured in amplitude and phase, correspond to dark points of vanishing electric field and undefined phase. Typically, a point of undefined phase in the transverse field structure occurs in the center of an optical (phase) vortex, i.e., an azimuthally increasing 2π-folded phase distribution [1]. The benefit of these singularities and respective topological structures have been proven, for instance, in optical micromanipulation, applied for, e.g., orbiting trapped objects [5, 6], or optical communication [21] with singularity-based integer information units. Beyond further applications, optical vortices can be considered as counterparts of vortices in hydrodynamics and matter waves, so that this analogy may be applied as a path to a deeper understanding and novel findings within these related areas [4].

Polarization singularities, i.e., vectorial singularities, are found in vectorial light of spatially varying polarization. Due to the vectorial character, there are different kinds of singularities which might occur, namely V-, C-, or L-point/ -line singularities [27, 31]. Within a polarization singularity a specific property of the corresponding polarization ellipse(s) is (are) undefined. Note that although, even naturally, space-varying polarization and vectorial singularities have obviously been there, there has been no particular interest in respective "exotic" light fields until the early 2000s [4]. At that time, vectorial singular light returned into the spotlight, when the intriguing focusing properties of radially and azimuthally polarized light fields were discovered [32, 33]. Since this discovery, research on structured vectorial light, its fundamentals and applications, has significantly increased. Nowadays, vectorial singular fields as cylindrical vector beams (CVBs) [34], representing an incoherent superposition of scalar helical Laguerre-Gaussian modes, are well-established in the structured light community. These as well as more tailored vectorial fields have given new insights into the singular properties of light [35–41] and have paved the way to advanced applications, e.g., in optical trapping [10, 11] and optical communication technologies [22, 23].

Considering applications, only recently, another originally quantum property of vectorial classical light has attracted growing attention: Specific vectorial fields as CVBs can be described as locally non-separable in their polarization and spatial degrees of freedom. This description is based on the fact that these singular fields are constituted of two scalar beams of orthogonal polarization (e.g., right- and left-handed circular polarization) and opposite helical phase structure. Due to their locally non-separable characteristic, these fields are also referred to as classically entangled [42]. Even though this notion is controversial [43], this classical analogon of local quantum entanglement has already proven its benefit for optical communication, enabling, e.g., higher-dimensional QKD [44] or the characterization of quantum channels by classical non-separable light [25]. This fusion of quantum and classical properties may open new perspectives for applied, but also fundamental optics.

Besides the transverse two-dimensional (2d) appearance of structured singular light, especially its behavior in three-dimensional (3d) space reveals interesting features. For instance, optical scalar and vectorial singularities may form optical links or knots, representing cutting-edge topological structures of singular optics and bearing the potential for advanced laser technology or optical trapping [45–49]. Furthermore, extended 3d micro- or nano-structures of light may contribute to the advancement of high-resolution imaging, optical micromanipulation, or material machining, among others. This ability has already been proven for scalar 3d structured fields, namely, non-diffracting and self-imaging beams (e.g., cf. Refs. [50–61]). In contrast, the study and effective implementation of vectorial 3d structured fields, tailored in amplitude, phase and, in particular, polarization, have received little attention so far, even though polarization would add a valuable and versatile degree of freedom to these kinds of fields. These vectorial fields would be of specific interest for working with, e.g., polarization sensitive materials. Until now, the respective required customization of vectorial structured light and especially its singularities in 3d space could not be realized so far.

One of the major current challenges in the field of structured light is the moving from structures of 2d to 3d polarization [4]. In general, paraxial light fields solely embed states of 2d polarization oscillating in the beams' transverse plane. For the realization of 3d polarization states, the moving to the non-paraxial regime can be the tool of choice. In this regime, non-negligible longitudinal in addition to transverse electric field contributions occur. First studies have been performed by different authors on 3d polarized fields, mainly concentrated on shaped intensity distribution in the tight focus of light [62–66]. In contrast, the corresponding focal (i.e., non-paraxial) 3d polarization structure is not yet studied that intensively. However, for future applications in, e.g., nano-material machining or nano-assembly of polarization sensitive materials, the miniaturization of shaped focal fields, thus, their nano-scale complexity, as well as the 3d polarization nature will be of huge benefit. As experimentally proven in 2015, particularly interesting 3d topologies are observed around polarization singularities within these non-paraxial fields, as optical Möbius strips [67, 68]. Being still in its infancy, until now, the potential of structuring the focal vectorial field including its singularities and 3d topologies is not even close to be fully exploited.

For the study and subsequent application of non-paraxial structured light the proper experimental characterization of realized fields represents the major challenge of today's research. With respect to the generation, various experimental approaches have been developed for the realization of structured paraxial fields. Focusing these fields aims at shaping non-paraxial light. A versatile tool for this purpose represents a spatial light modulator (SLM). It enables the formation of custom scalar light fields by computer-generated holograms [69], but also the indirect (interferometric) [31, 34, 70] or direct [36, 71–74] modification of the spatial polarization distribution. The combination of different techniques even allow for the customization of fields shaped in amplitude, phase, as well as polarization—*fully-structured* fields [74, 75]. Various different paraxial as well as non-paraxial fully-structured fields can be customized by named techniques. However, although the analysis of paraxial light is straightforward, the experimental characterization of non-paraxial fields is impeded by their nano-scale structure and, in particular, the 3d polarization nature. Therefore, their experimental study and effective implementation, required for the pending advancement of various different applications, is till now hindered.

In total, the combination of structured light and singular optics, i.e., structured singular light in the paraxial as well as non-paraxial regime, is poised to open entirely new opportunities in fundamental science as well as applied physics. The study of structured singular light has already given and will offer even more novel insights into the not yet fully explored nor understood nature of light and pave the way to further innovative applications in areas as optical micromanipulation, imaging, material machining, communication and many more. Many questions are still open or have not even yet been asked, especially considering polarization structuring—as cited above: "[...] there is always more to be learned [about light], just when the utmost depths seem within grasp." [1]. Therefore, this thesis concentrates on the study of the topical field of structured singular light. By customizing vectorial light in amplitude, phase and/ or polarization, cutting-edge questions on scalar and vectorial singularities are addressed, including their natural (in)stability, interaction and unfolding characteristics as well as the on-demand formation of multiple singularities in an individual paraxial or non-paraxial beam. Beyond, the control of topological field configurations is demonstrated, involving, e.g., fields fully-structured in 3d space, shaped in their focal intensity or 3d polarization topology. The benefit of these tailored singular fields, also including the property of quantum-like non-separability, for different applications will be discussed and proven within this thesis.

Comprising these and more topics, the thesis is structured to start from two-, going via three-, through to four-dimensionally (4d) structured singular light fields. Finally, properties of classical 3d fields are additionally fused with the quantum property of local entanglement or non-separability, enlarging the already rich phenomenology of structured light. In the following the corresponding four chapters will be outlined in more detail.

1.1 Chapter 2 | Fundamentals and Customization of Singular Light Fields

In order to build the base to understand and subsequently examine structured singular light fields, in Chap. 2, first, the required fundamentals are introduced. In the first section, singularities are elucidated as a general physical phenomenon including their properties and how to analyze them. Thereby, the characteristic of "genericity" is explained. The presented far-reaching approach is applied throughout the thesis to describe optical scalar and vectorial singularities in light. To interrelate between singularities and structured light, the fundamentals of tailored scalar and vectorial fields are presented subsequently, including the definition of respective singularities and their generic form in 2d or 3d structured paraxial fields of transverse (2d) polarization.

Second, chosen experimental SLM-based customization techniques, used within this thesis for the creation of tailored singular light, are presented. Revealing the not yet fully exploited potential of these techniques, the exemplary customization of innovative energy flow configurations and polarization structures including their singularities is proposed in 2d space. Resulting fields are analyzed with respect to their singular properties, showing the ability to use energy flow as a illustrative and instructive, easy interpretable tool to identify polarization singularities and giving new insights into the properties of vectorial singularities in very close vicinity. Finally a novel customization technique is proposed based on the spatial/ angular multiplexing of multiple holograms. The presented technique facilitates the modulation of fully-structured fields by a single hologram, thus, SLM. The cost-effective method is of interest for the advancement of set systems based on non-HD (HD: high definition) SLMs, which desire the addition of polarization structuring to a scalar modulation system.

1.2 Chapter 3 | Vectorial Light Fields and Singularities in 3d Space

Moving beyond structuring singular light in 2d space, in Chap. 3, vectorial complex light fields and respective singularities are considered in 3d space, i.e., upon propagation. Crucially, singular points or lines in 2d space form singular lines or surfaces in 3d space, respectively, enabling the creation of sophisticated 3d singularity configurations. However, till now, research has mainly been focused on fields in 2d transverse planes, scalar 3d structured field, or basic vectorial fields as CVBs in 3d space. In contrast, here, the 3d spatial customization of vectorial fields imparting scalar and/ or vectorial singularities is presented, revealing not yet known propagation dynamics of singularities and innovative beam classes.

In the frame of this chapter, a novel class of vectorial light fields is proposed, namely, Ince-Gaussian vector modes (IVMs). These modes are formed by the approach of incoherent superposition of scalar light fields, similar to CVBs. However, in contrast to similarly realized vectorial fields, these modes embed not an individual, but multiple singular V-points in its transverse 2d plane. Considering the application of polarization singularities as integer information units, this feature may be of benefit for the advancement of information technologies. In this applied area, the propagation of respective fields in 3d space is of particular interest. Therefore, the propagation dynamics of tailored IVMs and embedded singularities is studied numerically in free space and through turbulence, proving their value for future applications. Additionally, the presented study gives insights into the instability of non-generic V- and stability of generic C-points under perturbation, observing the transformation from the former to the latter.

In the next section, the study of singularity propagation dynamics is deepened by the investigation of combined non-generic, thus unstable, scalar phase and vectorial polarization singularities under minor perturbations within a joint tailored field. Although separately, these singularities have been studied intensively, their combination and, thus, their interaction in 3d space has not been under consideration so far. A numerical spectral method enables the in-depth investigation of this combination in a slightly perturbed system, forming singularity "explosions" in 3d space and revealing the not yet known ability to transfer unfolding characteristics from one singularity class to another. Further, an insight into the propagation dynamics of generic L- and non-generic C-singularities is given.

Next, it is shown that, not only the propagation dynamics of singularities can be affected by structured light techniques, but even the overall light field including its amplitude, phase and polarization can be tailored in 3d space. For this purpose, scalar light fields of sophisticated 3d propagation characteristics are chosen as bases. More precisely, non-diffracting (i.e., propagation invariant) as well as self-imaging scalar fields are used to form vectorial singular beams, enabling the introduction of novel beam classes. To experimentally shape these beams, two different approaches are applied both including a SLM as key component. Within this section, the novel vectorial 3d fields are studied numerically as well as experimentally, showing the ability to shape singularities and polarization in the transverse plane as well as longitudinally.

1.3 Chapter 4 | Non-paraxial 3d Polarization in 4d Light Fields

Till now, 2d and 3d vectorial paraxial light fields of 2d polarization were considered within this thesis. As indicated above, by tightly focusing radial polarization components, the formation of longitudinal in addition to transverse polarization contributions is enabled in the non-paraxial regime. Therefore, the creation of 3d polarization and, thus, 4d structured fields is facilitated, embedding the longitudinal electric field component as fourth dimension. These non-paraxial tailored fields are of specific benefit for the effective implementation of 4d materials, as stated in detail in Chap. 4, but are also required for the advancement of applications as imaging or optical micromanipulation.

To study these kind of fields and the ability to customize them on demand, first, the applied numerical approach is introduced. Subsequently, tailored non-paraxial intensity landscapes are shaped and analyzed in 2d and 3d space by tightly focusing phase and polarization modulated singular light fields. Here, singularity indices as the topological charge ℓ are proposed and analyzed as customization tools for focal fields including the ratio of longitudinal and transverse electric field contributions. The value of created light fields is subsequently proven by the experimental optical trapping and orienting of zeolite-L nano-containers.

As outlined above, the experimental identification of non-paraxial field properties represents a major topical issue, impeding the proper experimental study and application of 4d fields. Overcoming this problem, in the next section a path-breaking technology for the nano-tomography of 4d structured non-paraxial light is proposed, for which a functional 4d material is applied as sensor. The material represents a self-assembled monolayer, whose molecules react sensitive to amplitude, phase as well as 3d polarization of the exciting 4d field. Theoretically as well as experimentally, the innovative capabilities of this approach are presented, detecting typically invisible non-paraxial properties.

In the last section of this chapter, the ability to tailor 3d polarization singularities as well as topologies in non-paraxial fields is studied. As customization tool, phase vortices are added to higher-order vector fields, embedding a non-generic V-point. Besides numerically examining the feature of tailored 3d polarization singularities in 4d fields, topological structures are investigated around the optical axis as well as off-axis singularities. In addition to optical cones and ribbons novel topological structures formed by Möbius strips are evinced.

1.4 Chapter 5 | Entanglement in Classical Light

R. Boyd has once described structured light as a near-perfect test case for the study of the quantum-classical boundary [4]. In this chapter, this boundary will be studied by fusing valuable classical and quantum properties for scientific progress and advanced applications. Namely, classical tailored light fields will be examined with respect to their local non-separability or entanglement properties. For this purpose, first, the property of local entanglement in structured light and its characterization is introduced. In the next step, local entanglement is produced within sophisticated 3d structured fields, more precisely, in non-diffracting vector Bessel-Gaussian modes. As it has been proven in Chap. 3, these vectorial modes show self-healing characteristics, i.e., they self-reconstruct if being obstructed by an obstacle. Within this section, the relation between the level of self-healing and the degree of non-separability is studied experimentally. While the level of self-healing can be adapted by changing the size of the digitally generated obstacle, the degree of entanglement is quantified by the so-called vector quality factor and Bell parameter. Both evince the till now not known characteristic of self-healing of the degree of local entanglement, facilitated by shaping a 3d structured singular light field.

In the last section of this thesis, the originally quantum quantity of local entanglement is shown as spatially variant upon propagation of a tailored light field, even though the respective medium is considered as unitary. This curious finding is enabled by shaping a vectorial singular light field whose polarization varies upon propagation. The respective 3d field is created by counter-propagation of two orthogonally polarized CVBs. Theoretically as well as experimentally, it is proven that this light field spatially varies between being fully separable and non-separable, i.e., the degree of non-separability oscillates upon propagation. The experimental proof is performed by a sophisticated approach, combining holographic generation of vector modes and artificial digital propagation. Besides arising questions on the notion of local entanglement and offering an experimental method being of interest for, e.g., high-resolution imaging, a novel kind of spin-orbit (SO) interaction, namely, paraxial SO coupling in free space is found and proven.

References

1. Andrews DL, Babiker M (2012) The angular momentum of light. Cambridge University Press
2. Maiman TH (1960) Stimulated optical radiation in ruby. Nature 187:493
3. Eichler HJ, Eichler J, Lux O (2018) Lasers: basics, advances and applications, vol 220. Springer series in optical science. Springer
4. Rubinsztein-Dunlop H et al (2017) Roadmap on structured light. J Opt 19:013001
5. He H et al (1995) Direct observation of transfer of angular momentum to absorptive particles from a laser beam with a phase singularity. Phys Rev Lett 75:826
6. Friese M et al (1996) Optical angular-momentum transfer to trapped absorbing particles. Phys Rev A 54:1593
7. Woerdemann M et al (2013) Advanced optical trapping by complex beam shaping. Laser Photon Rev 7:839
8. Alpmann C, Schöler C, Denz C (2015) Elegant Gaussian beams for enhanced optical manipulation. Appl Phys Lett 106:241102
9. Andrews DL (2011) Structured light and its applications: an introduction to phasestructured beams and nanoscale optical forces. Academic
10. Zhan Q (2004) Trapping metallic Rayleigh particles with radial polarization. Opt Express 12:3377
11. Donato M et al (2012) Optical trapping of nanotubes with cylindrical vector beams. Opt Lett 37:3381
12. Hell SW, Wichmann J (1994) Breaking the diffraction resolution limit by stimulated emission: stimulated-emission-depletion fluorescence microscopy. Opt Lett 19:780
13. Török P, Munro P (2004) The use of Gauss-Laguerre vector beams in STED microscopy. Opt Express 12:3605
14. Biss DP, Youngworth KS, Brown TG (2006) Dark-field imaging with cylindricalvector beams. Appl Opt 45:470
15. Rose P et al (2013) Airy beam induced optical routing. Appl Phys Lett 102:101101
16. Meier M, Romano V, Feurer T (2007) Material processing with pulsed radially and azimuthally polarized laser radiation. Appl Phys A 86:329

17. Nivas JJ et al (2017) Surface structuring with polarization-singular femtosecond laser Beams generated by a q-plate. Sci Rep 7:42142
18. Rosales-Guzmán C et al (2013) Experimental detection of transverse particle movement with structured light. Sci Rep 3:2815
19. Töppel F et al (2014) Classical entanglement in polarization metrology. New J Phys 16:073019
20. Gibson G et al (2004) Free-space information transfer using light beams carrying orbital angular momentum. Opt Express 12:5448
21. Mirhosseini M et al (2015) High-dimensional quantum cryptography with twisted light. New J Phys 17:033033
22. Milione G et al (2015) Using the nonseparability of vector beams to encode information for optical communication. Opt Lett 40:4887
23. Milione G et al (2015) 4x 20 Gbit/s mode division multiplexing over free space using vector modes and a q-plate mode (de)multiplexer. Opt Lett 40:1980
24. Wang J (2016) Advances in communications using optical vortices. Photon Res 4:B14
25. Ndagano B et al (2017) Characterizing quantum channels with non-separable states of classical light. Nat Phys 13:397
26. Rosales-Guzmán C, Ndagano B, Forbes A (2018) A review of complex vector light fields and their applications. J Opt 20:123001
27. Dennis MR, O'Holleran K, Padgett MJ (2009) Singular optics: optical vortices and polarization singularities. In: Wolf E (ed) Progress in optics, vol 53, Chap. 5. Elsevier, pp 293–363
28. Soskin MS, Vasnetsov MV (2001) Singular optics. In: Wolf E (ed) Progress in optics, vol 42, Chap. 4. Elsevier, pp 219–277
29. Nye JF (1999) Natural focusing and fine structure of light: caustics and wave dislocations. CRC Pres
30. Berry MV, Dennis MR, Jr RLL (2004) Polarization singularities in the clear sky. New J Phys 6:162
31. Galvez EJ (2015) Light beams with spatially variable polarization. In: Photonics - fundamentals of photonics and physics, vol 1, Chap. 3. Wiley, pp 61–76
32. Quabis S et al (2000) Focusing light to a tighter spot. Opt Commun 179:1
33. Dorn R, Quabis S, Leuchs G (2003) Sharper focus for a radially polarized light beam. Phys Rev Lett 91:233901
34. Zhan Q (2009) Cylindrical vector beams: from mathematical concepts to applications. Adv Opt Photon 1:1
35. Cardano F et al (2013) Generation and dynamics of optical beams with polarization singularities. Opt Express 21:8815
36. Otte E, Alpmann C, Denz C (2016) Higher-order polarization singularitites in tailored vector beams. J Opt 18:074012
37. Lopez-Mago D et al (2013) Dynamics of polarization singularities in composite optical vortices. J Opt 15:044028
38. Flossmann F et al (2005) Polarization singularities from unfolding an optical vortex through a birefringent crystal. Phys Rev Lett 95:253901
39. Lu X et al (2014) Polarization singularities and orbital angular momentum sidebands from rotational symmetry broken by the Pockels effect. Sci Rep 4:4865
40. Vyas S, Kozawa Y, Sato S (2013) Polarization singularities in superposition of vector beams. Opt Express 21:8972
41. Ram BB, Sharma A, Senthilkumaran P (2017) Diffraction of V-point singularities through triangular apertures. Opt Express 25:
42. Aiello A et al (2015) Quantum- like nonseparable structures in optical beams. New J Phys 17:043024
43. Karimi E, Boyd RW (2015) Classical entanglement? Science 350:1172
44. Ndagano B et al (2017) A deterministic detector for vector vortex states. Sci Rep 7:13882
45. Berry MV, Dennis MR (2001) Knotted and linked phase singularities in monochromatic waves. Proc R Soc Lond A 457:2251

46. Berry MV, Dennis MR (2001) Knotting and unknotting of phase singularities: Helmholtz waves, paraxial waves and waves in 2+1 spacetime. J Phys A: Math Gen 34:8877
47. Leach J et al (2005) Vortex knots in light. New J Phys 7:55
48. Sugic D, Dennis MR (2018) Singular knot bundle in light. J Opt Soc Am A 35:1987
49. Larocque H et al (2018) Reconstructing the topology of optical polarization knots. Nat Phys 14:1079
50. Fahrbach FO, Simon P, Rohrbach A (2010) Microscopy with self-reconstructing beams. Nat Photon 4:780
51. Planchon TA et al (2011) Rapid three-dimensional isotropic imaging of living cells using Bessel beam plane illumination. Nat Methods 8:417
52. Fahrbach FO, Rohrbach A (2012) Propagation stability of self-reconstructing Bessel beams enables contrast-enhanced imaging in thick media. Nat Commun 3:632
53. Rogers G (1968) Fourier images in electron microscopy and their possible misinterpretation. J Microsc 89:121
54. Dubey AK, Yadava V (2008) Laser beam machining - a review. Int J Mach Tools Manuf 48:609
55. McGloin D, Dholakia K (2005) Bessel beams: Diffraction in a new light. Contemp Phys 46:15
56. McGloin D, Garcés-Chávez V, Dholakia K (2003) Interfering Bessel beams for optical micromanipulation. Opt Lett 28:657
57. Arlt J et al (2001) Optical micromanipulation using a Bessel light beam. Opt Commun 197:239
58. Volke-Sepulveda K et al (2002) Orbital angular momentum of a high-order Bessel light beam. J Opt B: Q Semiclass Opt 4:S82
59. Ulrich R, Kamiya T (1978) Resolution of self-images in planar optical waveguides. JOSA 68:583
60. Blume S (1976) Analogies between electronic and optical systems. Optik 46:333
61. Patorski K (1989) The self-imaging phenomenon and its applications. In: Wolf E (ed.) Progress in optics, vol 27. Elsevier pp 1–108
62. Otte E, Tekce K, Denz C (2017) Tailored intensity landscapes by tight focusing of singular vector beams. Opt Express 25:20194
63. Chen W, Zhan Q (2006) Three-dimensional focus shaping with cylindrical vector beams. Opt Commun 265:411
64. Bokor N, Davidson N (2006) Generation of a hollow dark spherical spot by 4pi focusing of a radially polarized Laguerre-Gaussian beam. Opt Lett 31:149
65. Wang H et al (2008) Creation of a needle of longitudinally polarized light in vacuum using binary optics. Nat Photon 2:501
66. Qin F et al (2015) Shaping a subwavelength needle with ultra-long focal length by focusing azimuthally polarized light. Sci Rep 5:9977
67. Freund I (2005) Cones, spirals, and Möbius strips, in elliptically polarized light. Opt Commun 249:7
68. Bauer T et al (2015) Observation of optical polarization Möbius strips'. Science 347:964
69. Davis JA et al (1999) Encoding amplitude information onto phase-only filters. Appl Opt 38:5004
70. Galvez EJ et al (2012) Poincaré-beam patterns produced by nonseparable superpositions of Laguerre-Gauss and polarization modes of light. Appl Opt 51:2925
71. Han W et al (2013) Vectorial optical field generator for the creation of arbitrarily complex fields. Opt Express 21:20692
72. Otte E et al (2015) Complex light fields enter a new dimension: holographic modulation of polarization in addition to amplitude and phase. Proc SPIE 9379:937908–937908
73. Alpmann C et al (2017) Dynamic modulation of Poincaré beams. Sci Rep 7:8076
74. Otte E, Tekce K, Denz C (2018) Spatial multiplexing for tailored fully-structured light. J Opt 20:105606
75. Gibson CJ et al (2018) Control of polarization rotation in nonlinear propagation of fully structured light. Phys Rev A 97:832

Chapter 2
Fundamentals and Customization of Singular Light Fields

The understanding and subsequent exploration of structured singular light fields is based on the fundamental knowledge of singularities and shaped light. To build this base, within this chapter, the required fundamentals and chosen customization techniques for structuring light are introduced. First, a general introduction on singularities and their properties (Sect. 2.1) is given. The presented approach will be applied throughout this thesis for the identification and characterization of all kinds of optical singularities. Subsequently, focusing on these singularities or, more precisely, singular light, the corresponding fundamental principles of light are outlined (Sects. 2.2 and 2.3). Thereby, the ability to shape light in its different degrees of freedom (DoFs), namely its amplitude, phase as well as polarization, is discussed. Furthermore, optical singularities appearing in scalar or vectorial structured fields, i.e., in fields sculpted in their amplitude and phase or (additionally) polarization, respectively, are introduced along with well-known exemplary fields. Subsequently, approaches for the customization of structured singular light are presented (Sect. 2.4). Besides elucidating established techniques, their still expandable capabilities are highlighted by sculpting and analyzing innovative polarization and energy flow configurations in the transverse plane, thus, two-dimensional (2d) space. Additionally, an innovative experimental approach is proposed for customizing singular light in amplitude, phase as well as polarization.

2.1 Singularities and Their Properties

Singularities represent an intriguing phenomenon that can be observed in various fields of research. While astrophysicists investigate gravitational or space-time singularities [1], material scientist are interested in dislocations in crystalline structures [2] and, in optics, optical singuarities in the intensity, phase and polarization of light are

E. Otte, *Structured Singular Light Fields*, Springer Theses,
https://doi.org/10.1007/978-3-030-63715-6_2

studied [3–5]. For all these cases, in the embedded singularity a specific property of the considered system cannot be defined. As a general approach, the system of interest can be described by a complex scalar function [6], namely

$$\Gamma = \nu + \mathrm{i}\mu = A \cdot \mathrm{e}^{\mathrm{i}\Phi}, \tag{2.1}$$

with the real ($\Re(\cdot)$) and imaginary ($\Im(\cdot)$) part $\nu = \Re(\Gamma)$ and $\mu = \Im(\Gamma)$, respectively, as well as the amplitude $A = |\Gamma|$ and phase $\Phi = \arg(\Gamma)$. By appropriate choice of parameters ν and μ, this representation is a valuable tool for the study of singularities. Within the system described by Γ, for the occurrence of a singularity of undefined characteristics, the real as well as imaginary part of the function needs to vanish, thus, $\nu = 0$ and $\mu = 0$. Hence, considering a 2d system, a singular point in Γ is identified by the intersection of zero-lines $Z_{\Re}$ and $Z_{\Im}$, for which $\nu = 0$ and $\mu = 0$, respectively. Typically, these point singularities are located in 2d vortex configurations of the respective phase Φ and can be characterized by a singularity index σ. Considering the surrounding Φ-structure, the singularity index or winding number $\sigma \in \mathbb{Z}$ can be derived [6] by the closed line integral

$$\sigma = \frac{1}{2\pi} \oint_C \mathrm{d}\Phi. \tag{2.2}$$

Here, C defines a closed non-selfintersecting loop around the singularity within Φ. The loop is directed positively, i.e., in counter-clockwise direction. Crucially, the singularity index represents a conserved quantity under perturbations of the considered system [7]. Further, a system cannot only include a single, but even multiple singular points, all identified as intersections of zero-lines. In this case, neighboring singularities on a joint zeroline reveal opposite index signs, following the so-called *sign rule* [7]. Singularities of the same sign can only be neighbors if a saddle point, i.e., an extremum is located between them on the zero-line. These extrema are marked as self-intersections of zero-lines [7].

Note that, besides appearing as a point, singularities may also be observed as line- or surface-singularities [6, 8]. For instance, if a point singularity is found in a 2d system (e.g., 2d space), in its 3d counterpart (e.g., 3d space) it will form a line. Accordingly, for line- or surface singularities in a system Γ, introduced zero-lines transform into zero-surfaces or even zero-volumes [4], respectively. For all kinds of singularities, one need to differentiate between natural and non-natural ones, whereby the former stay stable upon perturbation and the latter unfold into natural singularities. For this differentiation the concept of *genericity* is applied, being defined by two parameters [6]. The first one is the dimension of the control space in which the singularity is located. Within this thesis, the control space is equal to the 2d or 3d space in which the system is considered. The second parameter is the codimension of singularity—hence, the number of independent homogeneous conditions, which are required for the singularity's occurrence [6]. The difference of co- and spatial dimension reveals the singularities generic, thus, natural form with

$$\text{dimension of space} - \text{codimension} \begin{cases} < 0 \rightarrow & \nexists \text{ generic form,} \\ = 0 \rightarrow & \text{point singularity,} \\ = 1 \rightarrow & \text{line singularity,} \\ = 2 \rightarrow & \text{surface singularity.} \end{cases} \tag{2.3}$$

A generic singularity may occur naturally, but can of course also be shaped artificially, and is stable under perturbations [6]. In contrast, a non-generic (non-natural) singularity can only be realized theoretically, thus, in an idealized case, or artificially [6]. Non-generic singularities are unstable under minor perturbations,[1] i.e., they unfold into generic stable singularities if they were slightly perturbed. Note that the notion "generic" is not equal for all fields of research (e.g., cf. Refs. [9, 10]). "Generic" may also be understood as "typical" or "universal". However, within this thesis the outlined definition following Ref. [6] is applied.

As an illustrative example for the concrete application of the approach above, the complex number z_c is considered as a 2d singular system. Matching Eq. (2.1), the complex number and, thus, system is defined by the complex scalar function

$$z_c = x + \mathrm{i}y = r \cdot \mathrm{e}^{\mathrm{i}\varphi} \tag{2.4}$$

in 2d Cartesian (x, y) and polar (r, φ) coordinates [6]. The respective system is visualized in Fig. 2.1. Obviously, a singularity is found at the origin of φ, in which the polar angle is undefined and $r = 0$. In this point smooth lines of constant φ meet, so that a φ value cannot be defined. As typically observed, the structure surrounding the singularity in φ resembles a vortex, i.e., an azimuthal change of phase (2π folded).

For an undefined polar angle φ, r needs to be zero. Since for $x,\ y = 0$ follows $r = \sqrt{x^2 + y^2} = 0$ and $\tan\varphi = y/x$ is undefined, the x- and y-axis represent the zero-lines of the considered system (cf. white lines in Fig. 2.1b). Hence, the singularity is confirmed to be in the origin of the coordinate system. Considering Eq. (2.2), respective singular point in z_c can be characterized by an index of $\sigma = 1$.

To determine the generic form of z_c-singularities, smooth parameterization of the system is needed [6]. If the polar representation is chosen, only one condition, $r = 0$, is found for z_c or, more generally, Γ, being singular. The corresponding codimension would be one. However, the polar representation does not present a smooth parametrization for this problem since in addition to the condition of $r = 0$ the phase φ needs to be undefined, contradicting a codimension of one. In contrast, the Cartesian representation of the complex scalar field $\Gamma = z_c$ needs to chosen, which leads to the identification of two independent homogeneous conditions $x = 0$ and $y = 0$ for the singularity occurrence. Consequently, the codimension of this singularity is two. Thus, following Eq. (2.3), its generic form is a point in the 2d system or 2d space.

The introduced approach is a far-reaching one which can also be applied for the study of optical singularities in light, in the paraxial as well as non-paraxial regime [6,

[1] In the following, "(un)stable under minor perturbations" is partially abbreviated as "(un)stable", i.e., the additional comment "under minor perturbation" will be omitted for simplicity reasons.

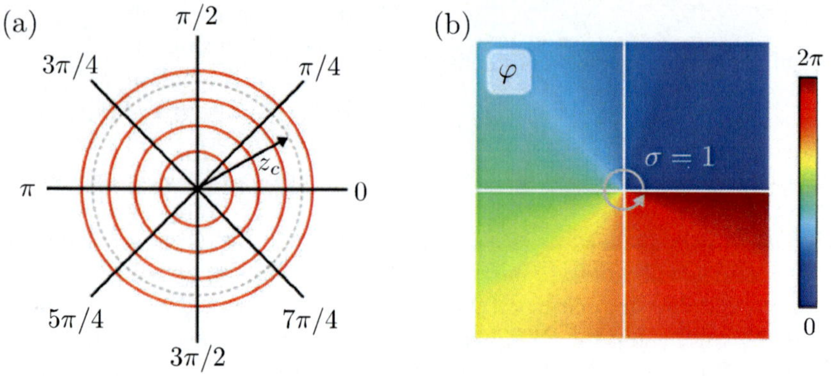

Fig. 2.1 Visualization of complex number z_c including a singularity of index $\sigma = 1$ in its origin. **a** z_c in the polar coordinate system (radii: red, polar angles: black), **b** zero-lines (white) in color-coded polar angle φ

11]. In the following, the fundamentals of these singular light fields are presented. Thereby the outlined concept for singularity identification and characterization will be applied for the introduction of singular properties in optical fields.

2.2 Fundamentals of Light

Singularities are not only a mathematical construct but can even be observed naturally in structured light [4]. Here, the fundamentals of these light fields are introduced, going from the full wave equation to singularities including their identification and genericity in scalar electromagnetic fields.

2.2.1 *From Full to Paraxial Wave Equation*

Light represents an electromagnetic wave being composed of the time t and position $\vec{r} = [x,\ y,\ z]^T$ (Cartesian coordinates) dependent electric and magnetic field vectors $\vec{\mathcal{E}}(\vec{r}, t)$ and $\vec{H}(\vec{r}, t)$. Applying Maxwell's equation of the vacuum, two identical wave equations for the electric and magnetic part can be derived [12] with

$$\left[\nabla^2 - \frac{1}{c^2}\frac{\partial^2}{\partial t^2}\right]\vec{\mathcal{E}}(\vec{r}, t) = 0, \qquad \left[\nabla^2 - \frac{1}{c^2}\frac{\partial^2}{\partial t^2}\right]\vec{H}(\vec{r}, t) = 0 \tag{2.5}$$

and the vacuum speed of light c. For simplicity, solely the electric field is considered in the following. In the case of a monochromatic harmonic wave oscillating with the angular frequency $\omega = 2\pi/T_p$ (T_p: period of harmonic oscillation), the separating ansatz

$$\vec{\mathcal{E}}(\vec{r}, t) = \Re\left[\vec{\mathcal{E}}(\vec{r})\, \mathrm{e}^{\mathrm{i}\omega t}\right] \qquad \text{with} \qquad \vec{\mathcal{E}}(\vec{r}) = \vec{e}(\vec{r})\mathcal{E}_0(\vec{r})\, \mathrm{e}^{\mathrm{i}\phi(\vec{r})} \tag{2.6}$$

is made [13]. The time dependent term is given by $\mathrm{e}^{\mathrm{i}\omega t}$, whereas all spatial dependent factors are included in $\vec{\mathcal{E}}(\vec{r})$ with $\vec{e}(\vec{r})$ being the polarization of light, $\mathcal{E}_0(\vec{r})$ the amplitude and $\phi(\vec{r})$ the phase. Applying this ansatz in Eq. (2.5) leads to the vectorial or full Helmholtz equation for the electric field given by

$$\nabla^2\vec{\mathcal{E}}(\vec{r}) + k^2\vec{\mathcal{E}}(\vec{r}) = 0 \tag{2.7}$$

with the wave number $k = |\vec{k}| = \omega/c = 2\pi/\lambda$ (wave vector $\vec{k} = [k_x, k_y, k_z]^T$) and the wavelength λ. Note that the complex electric field vector $\vec{\mathcal{E}}(\vec{r})$ represents a 2d or 3d vector dependent on the considered regime: in the non-paraxial regime the electric field vector is 3d with $\vec{\mathcal{E}}(\vec{r}) = [\mathcal{E}_x(\vec{r}), \mathcal{E}_y(\vec{r}), \mathcal{E}_z(\vec{r})]^T$, in the paraxial regime $\vec{\mathcal{E}}(\vec{r}) = [\mathcal{E}_x(\vec{r}), \mathcal{E}_y(\vec{r})]^T$ with $\mathcal{E}_z(\vec{r})$ being negligible. In the latter case $\vec{\mathcal{E}}(\vec{r})$ corresponds to the Jones vector [12] (cf. Sect. 2.3.1) for the polarization at position $\vec{r}$. For increasing ωt, $\vec{\mathcal{E}}(\vec{r}, t)$ describes an ellipse—the so-called polarization ellipse (cf. Sect. 2.3.1).

For both, the paraxial as well as the non-paraxial regime, the polarization of light may vary spatially. In the following, these beams are referred to as *vectorial light fields*. In contrast, if a light field of spatially homogeneous polarization $\vec{e}(\vec{r}) = \vec{e}\,\forall\,\vec{r}$ is assumed, *scalar light fields* are considered, simplifying the ansatz in Eq. (2.6) to the complex scalar wave

$$\mathcal{E}(\vec{r}) = \mathcal{E}_0(\vec{r})\, \mathrm{e}^{\mathrm{i}\phi(\vec{r})} \tag{2.8}$$

oscillating tangentially to the wave front. Hence, the vectorial or full Helmholtz equation reduces to its scalar version [14]

$$\nabla^2\mathcal{E}(\vec{r}) + k^2\mathcal{E}(\vec{r}) = 0. \tag{2.9}$$

By paraxial approximation with $\partial\mathcal{E}/\partial z \ll k\mathcal{E}$ and $\partial^2\mathcal{E}/\partial z^2 \ll k^2\mathcal{E}$ the paraxial scalar Helmholtz equation [15]

$$\left[\nabla_\perp^2 - \mathrm{i}2k\frac{\partial}{\partial z}\right]\mathcal{E}(\vec{r}) = 0 \tag{2.10}$$

is obtained with the transverse Laplacian operator $\nabla_\perp$. To easily differentiate paraxial from non-paraxial fields, paraxial light fields will be labeled as $\vec{E}(\vec{r}) = [E_x(\vec{r}), E_y(\vec{r})]^T$ or $E(\vec{r})$ for the vectorial or scalar case, respectively, in the following.

An elementary solution of paraxial scalar wave equation represents the fundamental Gaussian beam [16] derived as the lowest order eigenmode for the laser resonator

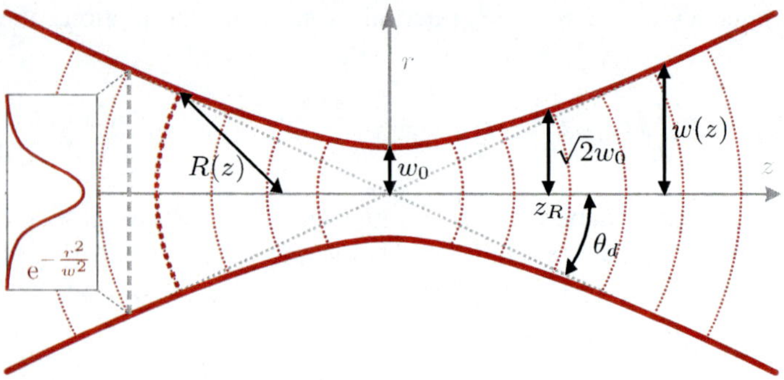

Fig. 2.2 Sketch of fundamental Gaussian beam in polar coordinates $(r,\ \varphi,\ z)$ with its beam waist $w_0 = w(z = 0)$, beam radius $w(z)$, Rayleigh length z_R, wave front curvature $R(z)$, and divergence angle θ_d. Inset depicts transverse Gaussian intensity profile

problem[2] (cf. Fig. 2.2). In polar coordinates $(r,\ \varphi,\ z)$, the respective complex electric field is defined as

$$E_G(\vec{r}) = E_{0,\max}\frac{w_0}{w(z)} \cdot \mathrm{e}^{-\frac{r^2}{w^2(z)}} \cdot \mathrm{e}^{-\mathrm{i}\left[kz+\frac{kr^2}{2R(z)}-\phi^G(z)\right]}. \tag{2.11}$$

The first exponential term reveals the Gaussian distribution of the amplitude dependent on radius r and with the maximal amplitude $E_{0,\max}$. Beyond, the function includes the beam waist $w_0 = w(z = 0)$, the beam radius $w(z) = w_0(1 + z^2/z_R^2)^{1/2}$ with the Rayleigh length $z_R = kw_0^2/2$, the wave front curvature $R(z) = z + z_R^2/z$, and the Gouy phase $\phi^G(z) = \arctan(z/z_R)$.

Besides this fundamental solution, several other exact solutions of the paraxial scalar wave equation (Eq. (2.10)) are known as Laguerre-, Hermite- or Ince-Gaussian modes, representing solutions in cylindrical, Cartesian or elliptical coordinates, respectively (cf. Sect. 2.2.3). Furthermore, as it will be introduced in Sect. 2.3.2, incoherent superpositions of these scalar solutions form vectorial light fields solving the vectorial wave equation in Eq. (2.7).

2.2.2 *Scalar Phase Singularities*

Comparing the complex scalar wave $E(\vec{r}) = \mathcal{E}(\vec{r})$ of Eq. (2.8) with the complex scalar function Γ in Eq. (2.1), it becomes evident that singularities in this electric field may occur—more precisely, scalar phase singularities in which the phase $\phi(\vec{r})$ is undefined and the amplitude $E_0(\vec{r}) = 0$. For the identification of these singularities

[2]Note that in many respects Bessel beams (more details in Sect. 3.3.1) are the most simple solutions with its transverse and longitudinal parts being completely factorized [13, 17, 18]. They represent solutions of the paraxial as well as full Helmholtz equation.

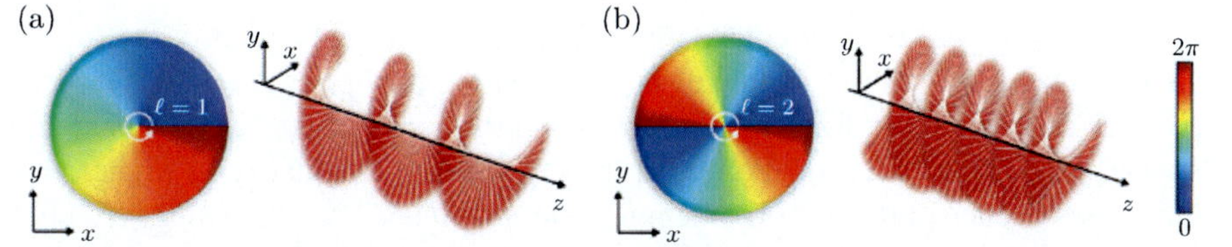

Fig. 2.3 Visualization of scalar phase singularity in 2d (left) and 3d (right) space. **a** Generic and **b** higher-order, non-generic scalar singularity of topological charge $\ell = 1$ and $\ell = 2$, respectively

the approach presented in Sect. 2.1 is applied. For this purpose, zero-lines $Z_{\Re}$ and $Z_{\Im}$ are chosen to correspond to $\nu = \Re(E) = 0$ and $\mu = \Im(E) = 0$ since

$$E(\vec{r}) = \nu + \mathrm{i}\mu = \Re(E(\vec{r})) + \mathrm{i}\Im(E(\vec{r})). \tag{2.12}$$

Hence, phase singularities are found if $Z_{\Re}$ and $Z_{\Im}$ cross each other, i.e., $\nu = 0$ as well as $\mu = 0$, and can therefore be seen as a two parameter problem [4, 19]. Since two independent conditions need to be fulfilled for the occurrence of a phase singularity, the codimension of this scalar singularity is two. Thus, following Eq. (2.3) its generic form is a point in 2d space and a line in 3d space [20–22], as illustrated in Fig. 2.3a.

Note that the parameters describing the investigated object, here, the scalar electric field E, need to be selected carefully as a smooth parameterization is needed [6]. Here, the Cartesian representation of the complex field E is chosen, which leads to the identification of two independent homogeneous conditions for the phase singularity occurrence. Similar to the example of complex number z_c in Sect. 2.1, the alternative polar representation does not present a smooth parametrization for this problem since in addition to the condition of $E_0 = 0$ the phase ϕ needs to be undefined. This fact contradicts a codimension of one, which would be identified for the polar representation ($E_0 = 0$). Consequently, the correct codimension of the scalar phase singularity is two.

Typically, scalar phase singularities are embedded in a phase vortex structure corresponding to a spatial twist of wave fronts and a helical trajectory of equal phase values in propagation direction z [16, 23–26]. Analyzing this topological structure according to Eq. (2.2), the respective singularity index, namely, the topological charge $\ell \in \mathbb{Z}$ of the phase singularity or vortex can be determined by

$$\ell = \frac{1}{2\pi} \oint_C \mathrm{d}\phi. \tag{2.13}$$

In the case of a generic phase singularity, i.e., for the required intersection of one $Z_{\Re}$- and one $Z_{\Im}$-line (codimension of 2), the singular point in 2d space possesses a topological charge of $|\ell| = 1$ (first-order phase singularity). However, higher-order non-generic phase vortices of charge $|\ell| > 1$ are well-established in the structured light community, e.g., for orbiting optically trapped particles or as information encoding basis [3, 25, 26]. An example of these structures is depicted in Fig. 2.3b with $\ell = 2$.

For $|\ell| > 1$ contour lines of constant phase values do not only cross each other, but also themselves. A similar behavior is observed for zero-lines $Z_{\Re}$ and $Z_{\Im}$. Generally, self-intersections of zero-lines correspond to saddle points, i.e., extrema in the studied field [7]. Hence, for the occurrence of a phase singularity of higher-order, the co-occurrence of $|\ell|$ first-order singularities and $|\ell| - 1$ saddle points is required, representing a codimension of $|\ell|(|\ell| + 1)$ [4, 27]. Consequently, higher-order phase singularities do not have a generic form in 2d or 3d space, thus, they are unstable in the non-idealized case. If they are artificially created as on-axis point singularities in 2d, under small perturbations a higher-order singularity will unfold into $|\ell|$ generic first-order phase singularities positioned on a circle surrounding the optical axis and twisting around the axis upon propagation [4, 28].

Note that there is an ongoing general discussion if non-generic optical singularities are realizable experimentally, since minor perturbations might be unavoidable. However, in the frame of this thesis the realization is assumed as possible with some restrictions. Details and studies on unfolding properties of non-generic singularities are given in Chap. 3.

2.2.3 Laguerre-Gaussian Modes

On the one hand, scalar phase singularities may appear naturally as, e.g., an ubiquitous phenomenon in interference, frequently occuring if light reflects from or propagates through random, turbulent, and chaotic media [4]. On the other hand, first- as well as higher-order phase vortices can be observed in man-made scalar light fields structured in amplitude and phase. The most famous example for these singular fields might be helical Laguerre-Gaussian (LG) modes. Originally, LG modes were derived as higher-order modes of the laser resonator problem [16, 29], representing an exact solution of the paraxial scalar Helmholtz Eq. (2.10) in polar coordinates (r, φ, z). Today, besides being generated in laser resonators, these fields are frequently realized by light shaping techniques as by spatial light modulators (SLMs; more details in Sect. 2.4.2).

These modes can be even (e), odd (o) or helical (h), whereby the latter ones represent the singular light fields of interest as they carry a helical phase front. The respective complex function with radial and azimuthal mode number $n \in \mathbb{N}_0$ and $\ell \in \mathbb{Z}$, respectively, is given by [29, 30]

$$\mathrm{LG}^h_{n,\ell}(r, \varphi, z) = A_{n,\ell}(r, z) \cdot \mathrm{e}^{\mathrm{i}\frac{kr^2}{2R(z)}} \cdot \mathrm{e}^{\mathrm{i}\phi^G_{n,\ell}(z)} \cdot \mathrm{e}^{\mathrm{i}\ell\varphi} \quad \text{with} \tag{2.14}$$

$$A_{n,\ell}(r, z) = \sqrt{\frac{2n!}{\pi(|\ell| + n)!}} \cdot \frac{1}{w(z)} \cdot \mathrm{e}^{-\frac{r}{w^2(z)}} \cdot \left(\frac{r\sqrt{2}}{w(z)}\right)^{|\ell|} \cdot \mathrm{L}_n^{|\ell|}\left(\frac{2r^2}{w^2(z)}\right), \tag{2.15}$$

$$\phi^G_{n,\ell}(z) = (2n + |\ell| + 1)\, \phi^G_{0,0}(z). \tag{2.16}$$

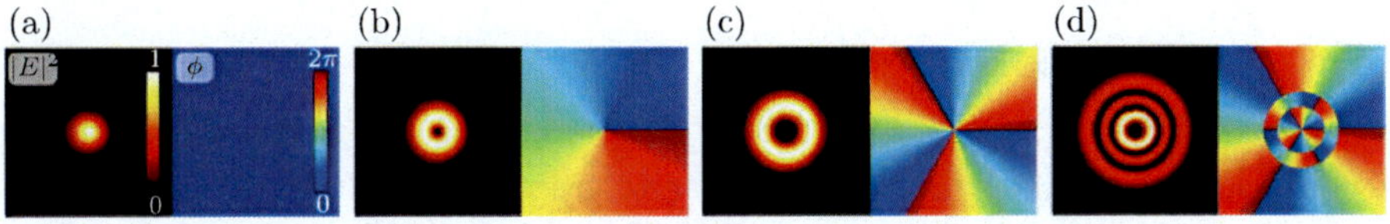

Fig. 2.4 Helical Laguerre-Gaussian $LG^h_{n,\ell}$ modes as examples for scalar singular light fields. Normalized transverse intensity $|E|^2$ and phase ϕ $(z = 0)$ are shown for **a** $LG^h_{0,0} = E_G$, **b** $LG^h_{0,1}$, **c** $LG^h_{0,-3}$, and **d** $LG^h_{2,3}$

Here, $L^{\ell}_n(\cdot)$ represents the eponymous Laguerre polynomial [31]. Equation (2.16) shows the Gouy phase shift $\phi^G_{n,\ell}(z)$ of LG modes. Obviously, the LG Gouy phase differs from the Gouy phase shift of the fundamental Gaussian modes $\phi^G(z) = \phi^G_{0,0}(z)$ dependent on mode indices n and ℓ. For $n,\ \ell = 0$, Eq. (2.14) reduces to the fundamental Gaussian mode (Eq. (2.11)).

Considering the occurrence of scalar phase singularities in these modes, the last exponential factor $\exp(i\ell\varphi)$ is of particular interest. This factor reveals the embedded phase vortex structure within a helical LG mode for $\ell \neq 0$. Thus, the respective LG light field includes an on-axis scalar phase singularity of charge ℓ, as exemplarily illustrated in Fig. 2.4 for different mode indices n and ℓ. LG modes represent exact self-similar solutions of the paraxial scalar wave equation, i.e., they maintain their transverse shape during propagation, even in the far field, only scaling due to divergence [29, 30]. In the frame of this thesis these light fields will be referred to as *structurally stable*.[3] Nevertheless, embedded perturbed singularities only propagate stably if they appear in their generic form,[4] i.e., if $|\ell| = 1$. However, due to their donut shaped intensity distribution with dark center, unfolding higher-order phase singularities do not disturb the structural appearance of propagating LG modes.

Besides contributing to the field of singular optics [4, 5], these beams are well-established tools in, e.g., optical communication or optical micromanipulation, in which scalar singularities with their index ℓ serve as integer information units [33–35] or the helical phase structure of LG modes is used to orbit trapped particle [36, 37], respectively. In the latter case, the optical angular momentum [23] or, more precisely, the *orbital angular momentum* (OAM) of these modes is applied. It is well-known that the momentum of light can be separated in its linear and angular part with the angular momentum being defined as [13]

---

[3]Consider that the notation "structural stability" is also used in singularity theory referring to the stability of singularities/critical points under small perturbations [32]. However, within this thesis this term labels the stability of the overall spatial structure of a light field, whereas for singularities the shortened term "stability" is applied.

[4]Note that if non-generic singularities are considered purely mathematically without any perturbations, hence in an idealized case, they may propagate stably, as visible from Eqs. (2.14)–(2.16). However, upon natural propagation including minor perturbations, i.e., in free space they would unfold.

$$\vec{J} = \varepsilon_0 \int \vec{r} \times (\vec{E} \times \vec{B})\, \mathrm{d}\vec{r} = \vec{L}_z + \vec{S}_z \tag{2.17}$$

with the magnetic induction $\vec{B} = \mu_0 \vec{H}$ and the free space electric (magnetic) permittivity ε_0 (μ_0). While the *spin angular momentum* (SAM) $\vec{S}_z$ corresponds to the polarization of light (cf. Sect. 2.3) with $\langle S_z \rangle = \pm\hbar$ per photon (positive/negative: left-/right-handed polarization), the OAM $\vec{L}_z$ depends on the spatial structure of the light field. For a phase vortex configuration as in LG modes, the OAM is associated with $\langle L_z \rangle = \ell\hbar$ per photon[5] [23, 39–41]. Both can be transferred to, e.g., optically trapped birefringent objects to spin or orbit these [36, 37]. Besides LG modes, there are more structured fields carrying OAM as, for example, non-diffracting higher-order Bessel beams [17, 18] (see Sect. 5.2.1). Generally, all these beams embedding a central phase singularity are often referred to as OAM modes of light [3, 26].

2.3 Polarization of Light

Besides beams being structured in amplitude and phase, referred to as scalar light fields, light spatially sculpted in polarization has returned to the spotlight [3]. These *vectorial light fields* can embed vectorial polarization singularities not only appearing in man-made fields but even representing natural phenomena in the blue daylight sky [4, 42]. Besides being part of fundamental research in singular optics [4] or in general wave phenomena, these fields have proven their benefit [3, 43] in, for instance, high-resolution imaging [44], information technologies [35, 45] or material machining [46]. In this section, the fundamentals of polarization are introduced leading to the outline of the characteristics of vectorial singular fields.

2.3.1 Formalism

2.3.1.1 Polarization Ellipse

The polarization $\vec{e}(\vec{r})$ of light defines the direction of oscillation of the general electric field vector $\vec{\mathcal{E}}(\vec{r}, t) = [\mathcal{E}_x(\vec{r}, t), \mathcal{E}_y(\vec{r}, t), \mathcal{E}_z(\vec{r}, t)]^T$ ($\vec{r} = [x, y, z]^T$; Eq. (2.6)). Here, for simplicity but without the loss of generality, the presented formalism is restricted to the paraxial regime with $\vec{\mathcal{E}}(\vec{r}, t) = \vec{E}(\vec{r}, t) = [E_x(\vec{r}, t), E_y(\vec{r}, t)]^T$ oscillating in the transverse (x, y)-plane, perpendicular to the z-direction of propagation. Moreover, the amplitude and phase are considered as spatially homogeneous, i.e., $E_0(\vec{r}) = E_0$ and $\phi(\vec{r}) = \phi$. Hence, the electric field vector $\vec{E}(\vec{r}) = \vec{E}$ is directly proportional to the state of polarization $\vec{e}(\vec{r}) = \vec{e}$, whereby with increasing time t or

[5] Note that this equation is not valid if more than one first- or higher-order phase singularity, thus, multiple phase vortices are located in a single light field [38].

ωt time dependent $\vec{E}(t) = \vec{E}\exp(\mathrm{i}\omega t)$ (cf. Eq. (2.6)) describes the respective *polarization ellipse* (cf. Sect. 2.2.1). Horizontally and vertically oriented contributions of the electric field $\vec{E}$ are represented by $E_x = E_{x,0}\exp(\mathrm{i}\phi_x)$ and $E_y = E_{y,0}\exp(\mathrm{i}\phi_y)$, respectively. Generally, a state of polarization is elliptically polarized. However, for specific amplitude differences $\Delta E_0 = |E_{x,0}| - |E_{y,0}|$ as well as relative phase differences $\Delta\phi = \phi_x - \phi_y$, the state of polarization can be linearly ($\Delta\phi = \{0, \pm\pi\}$) or circularly ($\Delta\phi = \pm\pi/2 \wedge \Delta E_0 = 0$) polarized.

If the complex electric field vector $\vec{E} = [E_x, E_y]^T$ is considered in the transverse (x, y)-plane, $\vec{E}$ can be written as

$$\vec{E} = \vec{p} + \mathrm{i}\vec{q} = (\vec{p}_0 + \mathrm{i}\vec{q}_0)\,\mathrm{e}^{\mathrm{i}\gamma} \tag{2.18}$$

with $\vec{p}_{(0)} = [p_{(0)x}, p_{(0)y}]^T$ and $\vec{q}_{(0)} = [q_{(0)x}, q_{(0)y}]^T$ being real 2d vectors and γ as so-called rectifying phase [11, 47] calculated form

$$\gamma = \frac{1}{2}\arctan\left(\frac{2\,\vec{p}\,\vec{q}}{p^2 - q^2}\right). \tag{2.19}$$

The rectifying phase is applied to go from non-orthogonal $\vec{p}$ and $\vec{q}$ to orthogonal vectors $\vec{p}_0$ and $\vec{q}_0$, representing the major and minor axis of the polarization ellipse, respectively (see Fig. 2.5a). Hence, the orientation Θ and ellipticity ε of the polarization ellipse is given by

$$\tan\Theta = \frac{p_{0y}}{p_{0x}}, \qquad \varepsilon = \frac{q_0}{p_0}. \tag{2.20}$$

Further, to determine the handedness of polarization the parameter

$$\begin{aligned} N^{\mathrm{p}} &= \Im(\vec{E}^* \times \vec{E}) \\ &= \vec{p} \times \vec{q} = \vec{p}_0 \times \vec{q}_0 \\ &= p_x q_y + p_y q_x \end{aligned} \tag{2.21}$$

can be applied.[6] For $N^{\mathrm{p}} > 0$ the state is right-handed, for $N^{\mathrm{p}} < 0$ it is left-handed polarized [11].

2.3.1.2 Müller Formalism and Stokes Vectors

As demonstrated above, in a transverse plane the electric field vector can be represented by a two component vector which corresponds to the Jones formalism. Here, the representation of states is limited to completely polarized light. However, if partially polarized or even unpolarized light shall be included, the Müller formalism is

[6] If the 3d vector $\vec{\mathcal{E}} = [\mathcal{E}_x, \mathcal{E}_y, \mathcal{E}_z]^T$ is considered in the non-paraxial regime, $\vec{N}^{\mathrm{p}} = \Im(\vec{E}^* \times \vec{E})$ is the 3d normal vector on the polarization ellipse.

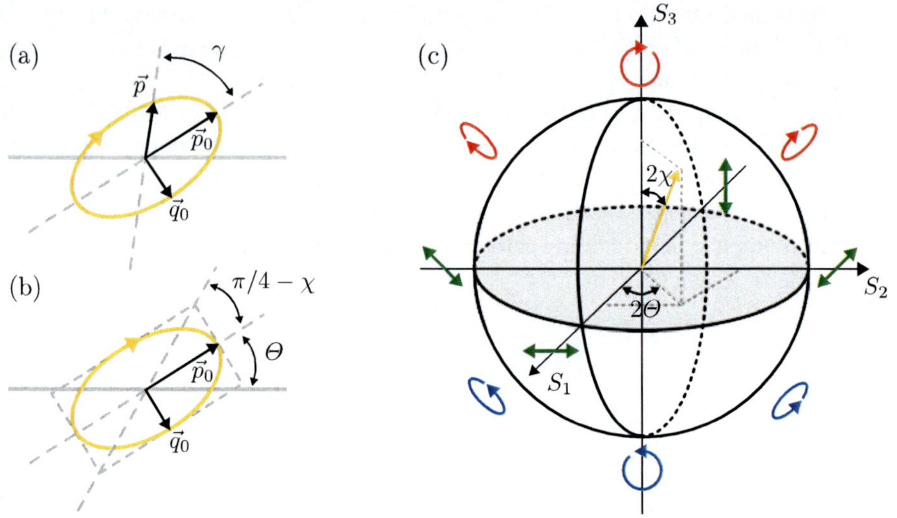

Fig. 2.5 Representation of polarization: Polarization ellipse **a** spanned by major axis $\vec{p}_0$ and minor axis $\vec{q}_0$ with rectifying phase γ, and of **b** ellipticity $\varepsilon = \tan(\pi/4 - \chi) = q_0/p_0$ and orientation Θ. **c** Poincaré unit sphere spanned by normalized Stokes parameters $S_{1,2,3}$ with linear states of polarization (green) located on its equator and circular states on its poles (north/south pole: right-/left-handed circular). Right-/left-handed elliptical states are found on the upper/lower hemisphere (red/blue)

applied [48]. In this case, states of polarization are described as a four component vector—the Stokes vector $\vec{s} = [s_0, s_1, s_2, s_3]^T$. To calculate changes in polarization if light passes through an optical system, both, Jones and Müller formalism can be applied. For both, calculations are based on matrices, whereby each optical component within the system is represented by a 2×2 or 4×4 matrix.

The parameters of the Stokes vector are calculated according to [49, 50]

$$\begin{aligned} s_0 &= |E_x|^2 + |E_y|^2, \\ s_1 &= |E_x|^2 - |E_y|^2, \\ s_2 &= 2\Re(E_x^* E_y), \\ s_3 &= 2\Im(E_x^* E_y). \end{aligned} \tag{2.22}$$

Here, $s_0 \in [0, 1]$ corresponds to the intensity of light, used for normalization of parameters according to $S_j = s_j/s_0$, $j = \{0, 1, 2, 3\}$ with $S_0^2 = 1 = S_1^2 + S_2^2 + S_3^2$. Further, $S_{1,2,3} \in [-1, 1]$ reflect the ratio of horizontal and vertical components, the ratio of diagonal and antidiagonal polarizations and the ellipticity of the state, respectively. Stokes parameters do not only serve as a mathematical description but can also be used for graphical visualization of polarization. For this purpose, normalized Stokes parameters are applied to span the Poincaré unit sphere [12], as illustrated in Fig. 2.5c. Each point on or within this sphere belongs to another state of polarization, varying in orientation, ellipticity and degree of polarization. Fully polarized states are located on the surface of this sphere, whereas completely unpolarized light is found in its origin. The upper and lower hemisphere corresponds to right-handed (red) and left-handed (blue) states of polarization, respectively, with pure circular polarization at the poles. Linear states are positioned on the equator of the Poincaré sphere. Note that each point on the sphere is directly related to the excentricity χ or ellipticity ε and orientation Θ of respective polarization ellipse with [51]

$$\varepsilon = \tan(\pi/4 - \chi), \qquad \chi = \frac{1}{2}\arcsin(S_3),$$
$$\Theta = \frac{1}{2}\arctan\left(\frac{S_2}{S_1}\right). \tag{2.23}$$

Points on opposite sites of the sphere correspond to orthogonal states of polarization as, e.g., right-circular and left-circular polarization located at the north and south pole, respectively.

Besides being applied as a tool for visualization of single polarization states, the Poincaré sphere has proven its benefit if light fields of inhomogeneous polarization are studied. For example so-called full-Poincaré beams [52] bear their name since in their transverse plane they include all states of polarization found on the surface of the Poincaré sphere. In contrast *vector fields* only embed states located on the equator of the sphere [51], i.e., linear states (more details in Sect. 2.3.2). Beyond, the concept of this unit sphere was extended to "Higher-order Poincaré spheres" (HOPSs) representing scalar and/or vectorial light fields, whereby opposite points correspond to orthogonal spatial modes [53]. In the following light fields of inhomogeneous polarization—vectorial light fields—are introduced in detail.

2.3.2 Vectorial Light Fields and Polarization Singularities

Structured polarization as well as respective vectorial polarization singularities may occur naturally [4, 42] as well as artificially. Considering the latter, similar to scalar light fields, polarization structured fields, i.e., *vectorial light fields* can be created by different modulation techniques based on interference, birefringence or holography [3, 54–57] (see Sect. 2.4). Generally, it can be differentiated between *vector fields*,[7] embedding only linearly polarized states, and *ellipse fields*, which may include linear as well as circular and elliptical states. Both may include vectorial polarization singularities, as it will be introduced in the following.

[7]Consider that the term "vector field" is not used equally to "vectorial (light) field". While the latter one includes all kinds of polarization structured light fields, the former one only represents fields of pure linear polarization. However, note that throughout literature the nomenclature varies.

2.3.2.1 Vectorial Singularities and Complex Stokes Fields

Due to the vectorial nature of polarization, there is a broad range of different generic and non-generic singularities occuring in vectorial light fields. In general, a vectorial singularity is found if at least one parameter of a polarization ellipse, i.e., the orientation Θ, handedness/ellipticity ε or even the complete ellipse itself, is undefined [51]. If the polarization ellipse, i.e., the electric field $\vec{E}$ cannot be defined, a V-point is found. For undefined orientation or handedness a C-singularity of pure circular polarization or L-singularity of pure linear polarization, respectively, is observed. Both, C- and L-singularities exist as point and line singularity in 2d space. However, there is only one generic form of these singularities in 2d as well as in 3d space, which will be discussed in the following.

In order to identify and analyze vectorial singularities in polarization structures, complex Stokes fields [58, 59] represent an established tool. Following the approach presented in Sect. 2.1, Stokes fields are defined as complex scalar fields (cf. Eq. (2.1)) constituted of the normalized Stokes parameters with

$$\Sigma_{ij} = S_i + \mathrm{i}S_j = A_{ij}\,\mathrm{e}^{\mathrm{i}\Phi_{ij}}, \qquad i, j = \{1, 2, 3\}. \tag{2.24}$$

The amplitude is given by $A_{ij} = (S_i^2 + S_j^2)^{1/2}$ and the complex Stokes field's phase by $\Phi_{12} = \arg(\Sigma_{ij})$. Similar to scalar phase singularities, in 2d space vectorial point singularities can be identified as phase singularities in Φ_{ij}-vortices with $A_{ij} = 0$ and respective Stokes singularity index

$$\sigma_{ij} = \frac{1}{2\pi}\oint_C \mathrm{d}\Phi_{ij}. \tag{2.25}$$

In the respective point the Stokes vector or the relation between individual Stokes parameters is undefined, i.e., at least one parameter of the respective polarization ellipse is not specified. Employing zero-lines $Z_{\Re}$ and $Z_{\Im}$ with $S_i = 0$ and $S_j = 0$, respectively, vectorial point singularities correspond to intersections, saddle points to self-intersections of these lines. As Stokes parameters represent a smooth parametrization, they can be used to determine the codimension and, thus, the generic form of vectorial singularities [6].

Note that besides being used for the identification of point singularities, Stokes fields may reveal line singularities in 2d space as presented in Ref. [28]. Moreover, investigating the Stokes fields in various 2d planes of a 3d volume enable the analysis of singularities in 3d space.

2.3.2.2 Vector Fields

Most established vector fields represent radial and azimuthally polarized beams being constituted of linearly polarized states pointing radially to the optical axis or being oriented azimuthally around this axis [43, 60, 61]. Due to the orientation of embed-

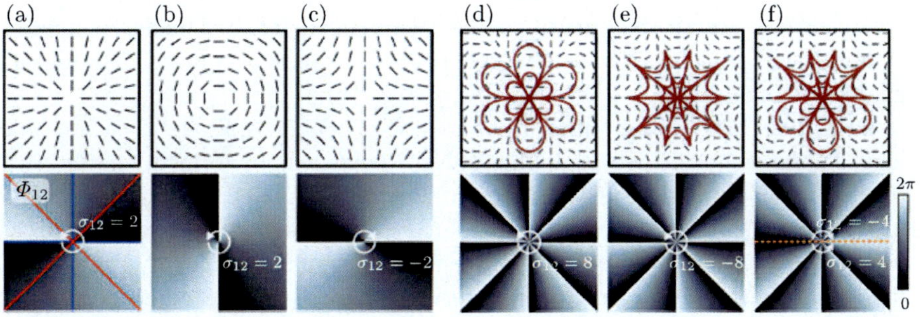

Fig. 2.6 Non-generic V-points in 2d vector fields (numerics): The polarization structure (top) is indicated by black lines with its respective Stokes field phase Φ_{12} shown underneath (transverse plane). **a–c** First-order vector fields with $|\sigma_{12}| = 2$ and zero-lines $Z_{\Re}$ (red) and $Z_{\Im}$ (blue). **d** Flower- ($\sigma_{12} = 8$) and **e** spider-web-shaped ($\sigma_{12} = -8$) higher-order vector field with red flow lines. **f** Hybrid structure of flower and web embedding two singularities is very close vicinity ($\sigma_{12} = \pm 4$)

ded states of polarization, considered in 2d space, a V-point of undefined polarization occurs on-axis (idealized case). Figure 2.6a, b depict the respective polarization distribution (top) by black lines indicating the polarization vector and phase Φ_{12} (bottom) in the beams' transverse plane. Analyzing the polarization vectors around the optical axis reveals all states of polarization located on the equator of the Poincaré sphere. Further, the vectors rotate by 360° (2π, counterclockwise) within a full counterclockwise circle (2π) around the singularity. The ratio of this rotation angle and the 2π circle defines the Poincaré-Hopf index $\eta_{\mathrm{PH}} \in \mathbb{Z}$ of the embedded singularity [32, 58]. Here, this index is $\eta_{\mathrm{PH}} = 1$, explaining why these beams are referred to as first-order vector fields with first-order V-point. Of course there are more first-order modes with $|\eta_{\mathrm{PH}}| = 1$, one of these is presented in Fig. 2.6c, showing a $-360°$ rotation of vectors. Note that a V-point only occurs for $\eta_{\mathrm{PH}} \neq 0$.

To study vector fields by introduced complex Stokes fields, Stokes parameters S_1 and S_2 are applied to constitute Σ_{12} and Φ_{12}. All states of polarization are linear, $S_3 = 0$, and thus, the handedness of polarization is undefined within the whole transverse plane.[8] Consequently, here a V-point corresponds to a point of undefined orientation being related to $S_{1,2}$ and represented by a singular point in Φ_{12}. The respective zero-lines $Z_{\Re,\Im}$ are formed for $S_{1,2} = 0$, as indicated in Fig. 2.6a (red, blue line). For fields in Fig. 2.6a–c, the (a) intersection of $Z_{\Re}$ and $Z_{\Im}$ and the (a)–(c) phase structure of Φ_{12} reveal on-axis first-order V-points of index $\sigma_{12} = \pm 2 = 2\eta_{\mathrm{PH}}$ [58]. For the occurrence of a V-point multiple $Z_{\Re}$ and $Z_{\Im}$ lines need to cross, evincing a codimension of four for a first-order V-point.[9] Hence, in both, 2d and 3d space V-points represent non-generic, thus, unstable vectorial singularities (cf. Eq. (2.3)),

[8] Since the whole transverse plane of this light field is undefined with respect to its handedness in polarization, this does not represent a vectorial C-singularity.

[9] The codimension of four of a first-order V-point is given by the conditions $\Re(E_x) = 0, \Re(E_y) = 0$, $\Im(E_x) = 0$, and $\Im(E_y) = 0$ resulting in $\vec{E}$ being undefined.

not occurring naturally but in artificially or theoretically created fields. As it will be demonstrated in Chap. 3, under small perturbations these singularities unfold into generic singular points.

Even though embedded singularities are non-generic, besides fields of first-order, higher-order vector fields with higher-order V-points can be formed with $|\sigma_{12}| > 2$ ($|\eta_{PH}| > 1$, codimension > 4). Some numerical examples with respective Stokes field phase Φ_{12} are visualized in Fig. 2.6d–f. Typically, higher-order vector fields can be separated according to the sign of respective singularity index σ_{12}: for positive indices, a vectorial *flower* structure is observed, whereas negative indices corresponds to vectorial *spider web* configurations [62, 63]. Here, included states of polarization form $|\sigma_{12} - 2|$ petals or sectors, respectively, as indicated in Fig. 2.6d, e for $\sigma_{12} = \pm 8$. Beyond, modern structured light techniques (cf. Sect. 2.4) allow for the realization of so-called *hybrid* structures [62], which are composed of flower petals as well as spider web sectors, as shown in Fig. 2.6(f). Interestingly, these fields embed a positive as well as negative polarization singularity in very close vicinity. Due to their closeness, the rule for the number of petal/sectors does not hold here, but is recovered if the distance of singularities is increased [62] as it will be demonstrated in Sect. 2.4.3. For characterization of hybrid structures a total singularity index $\sigma_{12}^{\mathrm{tot}} = \sum_v \sigma_{12,v}$ is defined with v referring to the number of embedded singularities. For Fig. 2.6(f), $\sigma_{12}^{\mathrm{tot}} = +4 - 4 = 0$ with the positive and negative singularity annihilating each other.

2.3.2.3 Ellipse Fields

Polarization structured fields including states of different ellipticity, i.e., S_3 varies spatially, are referred to as ellipse fields. In contrast to vector fields, these beams can include not only V-points but also C- and L-singularities. An example of an ellipse field (numerics) including an on-axis V-point is illustrated in Fig. 2.7a (top) with red/blue ellipses corresponding to right-/left-handed states and green lines to linear polarization. Obviously, a V-point occurs if all states of polarization surrounding a point annihilate each other. As a consequence, for the occurence of a V-point in an ellipse field C- (gray lines) and L-lines (green lines), i.e., lines of pure circular and linear polarization, respectively, need to cross. Note that ellipse fields embedding a V-points are also known as "Poincaré vortices" as they are analyzed by the Stokes field $\Sigma_{23} = S_2 + \mathrm{i}S_3$ referred to as "Poincaré field"[10] named after H. Poincaré himself [28, 64]. Here, a V-point is found as phase singularity in Φ_{23} (Fig. 2.7a, bottom) and characterized by singularity index σ_{23} with $|\sigma_{23}| = 2$ or $|\sigma_{23}| > 2$ for lowest- or higher-order V-points, respectively.

However, neither the V-point nor the C-line represent a generic singularity. Similar to vector fields, the lowest-order V-point in an ellipse field has a codimesion of four, corresponding to multiple zero-lines $Z_{\Re}$ and $Z_{\Im}$, here each with $S_2 = 0$ and $S_3 = 0$,

[10] In the presented case, $S_1 = 0$. However, for the formation of a V-point in an ellipse field, Stokes parameters could also be $S_2 = 0$ and $S_{1,3}$ being spatially variant within the transverse plane. In this case, Stokes field $\Sigma_{31} = S_3 + \mathrm{i}S_1$ is analyzed and zero-lines are constituted of $S_3 = 0$ and $S_1 = 0$.

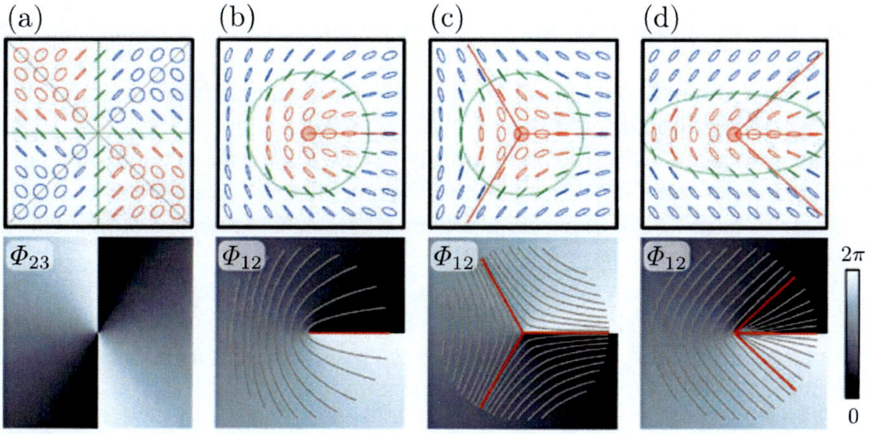

Fig. 2.7 Vectorial singularities in exemplary ellipse fields (numerics). Polarization distribution (top) embedding linear (green), right- (red) and left-handed (blue) elliptical/circular states of polarization with respective Stokes field phase $\Phi_{12,23}$ (bottom). **a** Poincaré vortex with C- (gray) and L-lines (green) crossing in the on-axis V-point ($\sigma_{23} = -2$). **b–d** Generic lowest-order C-points in **b** lemon ($\sigma_{12} = 1$), **c** star ($\sigma_{12} = -1$), or **d** monstar ($\sigma_{12} = 1$) configuration with red ζ-lines and gray flow lines

respectively, intersecting in this point. Thus, following Eq. (2.3) the V-point is non-generic in 2d and 3d space. Moreover, a singularity of pure circular polarization, i.e., a C-singularity ($S_3 = \pm 1$) requires undefined orientation of considered states, thus, $S_1 = 0$ and $S_2 = 0$ (two conditions, thus, codimension = 2). Thus, the generic form of a C-singularity in 2d/3d is a point/line, i.e., a C-line in 2d is non-generic. Hence, V-points and C-lines unfold under minor perturbation. The only generic singularity observed in the ellipse field in Fig. 2.7a is the L-line: L-singularities have a codimension of one with the requirement of $S_3 = 0$, hence, their generic, stable form in the paraxial regime is a line/surface in 2d/3d space.

Fig. 2.7b–d show examples of ellipse fields embedding generic C-point and L-line singularities. With its conditions $S_1 = 0$ and $S_2 = 0$, C-points are identifiable as singular points in phase Φ_{12} of Stokes field Σ_{12}, i.e., as intersections of zero-lines $Z_{\mathfrak{R}}$ with $S_1 = 0$ and $Z_{\mathfrak{I}}$ with $S_2 = 0$. Generic C-points reveal indices of $|\sigma_{12}| = 1$ but can also be characterized by the index $I_C = \frac{1}{2\pi} \oint_C \mathrm{d}\Theta = \sigma_{12}/2$ studying the orientation of the major axes around the singular point [11, 58]. Since the major axis represents a headless vector and is consequently π-invariant under rotation [65], the smallest possible index is $|I_C| = 1/2$ [11] corresponding to lowest-order C-points ($|\sigma_{12}| = 1$). Depending on the rotation direction of surrounding states, there is one (Fig. 2.7b) or there are three (Fig. 2.7c, d) so-called ζ-lines appearing along which the states' major axes are oriented radially to these C-points. Following the number of ζ-lines and the index $I_C = \sigma_{12}/2$, the structures for lowest-order C-points can be separated into *lemon*, *star* and *monstar* topologies [66, 67]. A lemon structure appears for $I_C = 1/2$ ($\sigma_{12} = 1$) with one ζ-line, as visualized in Fig. 2.7b. Here, the red line in the polarization distribution as well as phase Φ_{12} represent the respective

ζ-line with additional gray flow lines in Φ_{12}. Similarly, Fig. 2.7c presents a star configuration with three ζ-lines and an index of $I_C = -1/2$ ($\sigma_{12} = -1$). As indicated by its name, monstars include properties of both, lemon and star structures: Here, three ζ-lines are detected whereby the index is positive with $I_C = 1/2$ ($\sigma_{12} = 1$), as exemplarily shown in Fig. 2.7d. Additionally, as the ellipticity ε of polarization needs to decrease/increase with distance from a C-point, there is typically a closed generic L-line ($\varepsilon = 0$, i.e., $S_3 = 0$) observed around a generic C-point separating right- and left-handed elliptical states (green line in Fig. 2.7b–d).

Note that besides generic lowest-order C-points, higher-order ones with $|\sigma_{12}| > 1$ can be realized artificially or theoretically. Equivalent to higher-order V-points these singular points require multiple zero-lines to cross, representing the non-genericity and, thus, instability of higher-order C-points.

2.3.2.4 Cylindrical Vector Beams and Full-Poincaré Modes

Originally, polarization structured light fields as radially or azimuthally polarized vector fields, theoretically embedding V-points, were derived as incoherent superpositions of spatial scalar modes of light. The most famous class of these fields might be cylindrical vector beams (CVBs) [43]. These beams represent exact analytical solutions of the vectorial Helmholtz equation (Eq. (2.7)), as they are constituted of incoherently superimposed exact solutions of the scalar paraxial wave equation (Eq. (2.10)) in cylindrical (polar) coordinates, namely LG modes (see Sect. 2.2.3).[11] Superimposed modes are of orthogonal spatial structure, i.e., opposite topological charge ℓ and the same radial index n, and orthogonal polarization. For the latter, e.g., circular, linear or diagonal polarization basis can be chosen [51]. The respective electric field is calculated from

$$\vec{E}_{\mathrm{CVB}} = \frac{1}{\sqrt{2}}\left[\vec{e}_1 \cdot \mathrm{LG}^{h,1}_{n,\ell} \cdot \mathrm{e}^{\mathrm{i}\alpha} + \vec{e}_2 \cdot \mathrm{LG}^{h,2}_{n,-\ell} \cdot \mathrm{e}^{-\mathrm{i}\alpha}\right] \tag{2.26}$$

with unit vectors $\vec{e}_{1,2}$ representing orthogonal states of polarization and α the phase relation between spatial modes. Figure 2.8a–f depicts some well-known examples of CVBs (numerics) realized by circular polarization basis (right-circular: $\vec{e}_1 = \vec{e}_R = [1,\ -\mathrm{i}]^T/\sqrt{2}$, left-circular: $\vec{e}_2 = \vec{e}_L = [1,\ \mathrm{i}]^T/\sqrt{2}$).[12] Beyond, if one of the spatial modes is replaced by a fundamental Gaussian beam, e.g., as $\mathrm{LG}^{h,2}_{n,\ell} = \mathrm{LG}^{h,2}_{0,0}$, while $\mathrm{LG}^{h,1}_{n,\ell}$ exhibits $n = 0$ and $\ell = 1$, a vectorial field embedding a generic C-point will be created as examplified in Fig. 2.8g, h. By this approach so-called full-Poincaré modes can be realized which contain all states of polarization of the Poincaré sphere within

[11]Even though cylindrical vector modes are the most established solutions, a broad range of more modes is accessible on the basis of different scalar paraxial solutions as, e.g., Hermite-Gaussian, Ince-Gaussian, Bessel or even Mathieu beams [51, 68–71].

[12]Note that the same structures can be formed, e.g., in linear polarization basis (e.g., $\vec{e}_x = [1,\ 0]^T$, $\vec{e}_y = [0,\ 1]^T$) by superposition of even (e) and odd (o) LG modes, which are constituted of two orthogonal helical LG beams (same polarization).

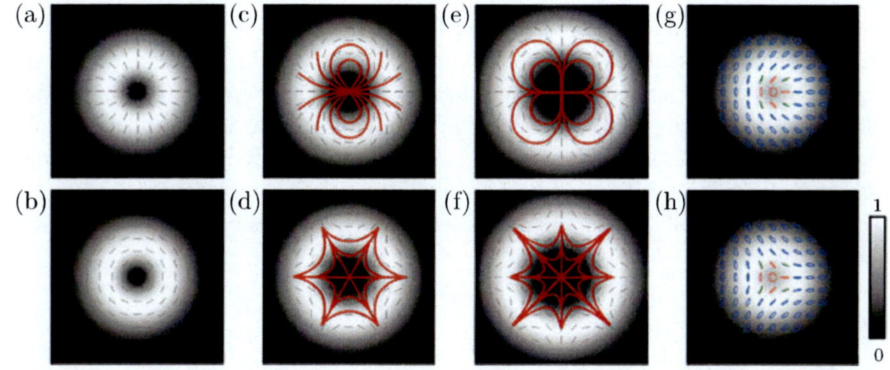

Fig. 2.8 Examples of vectorial field realized by incoherent superposition of spatial modes (numerics). **a–f** Cylindrical vector beams (CVBs; red flow lines) formed by superposition of scalar helical LG modes with $n = 0$ and **a** $\ell = 1, \alpha = 0$, **b** $\ell = 1, \alpha = \pi/2$, **c** $\ell = 2, \alpha = 0$, **d** $\ell = -2, \alpha = 0$, **e** $\ell = 3, \alpha = 0$, or **f** $\ell = -3, \alpha = 0$. **g, h** Generic C-point formation by superposition of fundamental Gaussian mode ($\mathrm{LG}_{0,0}^{h,2}$) and LG mode ($\mathrm{LG}_{0,\ell}^{h,1}$) with **g** $\ell = 1$ or **h** $\ell = -1$

their transverse plane [52]. Since full-Poincaré modes and CVBs are structured in amplitude, phase, as well as polarization, these beams and equivalent ones are referred to as *fully-structured* fields [72].

Note that, following superposition principle, CVBs are considered as structurally stable, i.e., their transverse appearance does not change upon propagation aside from enlarging in size due to diffraction. However, included V-points (a)–(f) are still unstable due to their codimension larger or equal to four, hence, unfold upon perturbation.[13] Obviously, V-points are located in the dark central area of the vectorial spatial mode. Hence, even though they unfold, this effect cannot be observed in CVBs and does not affect the transverse polarization distribution within the donut shaped intensity.

2.4 Experimental Customization of Singular Light

Within the last decades various methods have been developed for the realization of singular light fields structured in amplitude, phase as well as polarization (e.g., see Refs. [56, 62, 73–80]). For this purpose spatial light modulars (SLMs) have emerged as a powerful tool as they enable a digital and dynamic variation of tailored spatial configurations. On the one hand, weighted blazed gratings allow for the realization of amplitude and phase modulated scalar light fields [73–75]. On the other hand, in combination with an interferometric system [57, 78, 79, 79, 81–85] or selected wave plates [56, 62, 76, 80, 86] the on-demand formation of polarization structures

[13]Considered purely mathematically without any perturbations, hence in an idealized case, non-generic V-points do not unfold upon propagation, as visible from Eqs. (2.14)–(2.16) and (2.26).

as well as of fully-structured fields is facilitated. In the following, selected methods are introduced for sculpting light based on SLMs. For established techniques, their capability is demonstrated, e.g., by sculpting sophisticated irregular vector fields (Sect. 2.4.3) or shaping innovative energy flow configurations in tailored polarization structures (Sect. 2.4.2) in 2d space. Moreover, a novel approach for fully-structuring light is proposed (Sect. 2.4.4), based on spatial/angular multiplexing.

Note that the results presented in this section have in parts been published in Refs. [28, 62, 86, 87].

2.4.1 Amplitude and Phase Structuring

Although amplitude modulation is required in addition to phase structuring, a phase-only SLM is an established tool for the realization of scalar shaped light fields. Generally, phase structures can easily be encoded digitally by computer generated holograms. Following the ansatz of J. A. Davis et al. [73] for the addition of amplitude modulation, the SLM function is assumed to be

$$T = \exp(\mathrm{i}E_0\phi) \tag{2.27}$$

for encoding the scalar light field $E(x, y) = E_0(x, y) \cdot \exp(\mathrm{i}\phi(x, y))$ (normalized amplitude $E_0 = E_0(x, y) \in [0, 1]$, phase $\phi = \phi(x, y) \in [0, 2\pi[$). Expressed as a mixed Fourier-Taylor series Eq. (2.27) is given by

$$T = \sum_{\nu=-\infty}^{\infty} T_\nu \mathrm{e}^{\mathrm{i}\nu\phi} \quad \text{with} \quad T_\nu = \mathrm{e}^{-\mathrm{i}(\nu - E_0)\pi} \,\mathrm{sinc}(\nu - E_0). \tag{2.28}$$

Here, ν is the diffraction order and $\mathrm{sinc}(*) = \sin(\pi \cdot *)/(\pi \cdot *)$. If $E_0 = 1$, the first diffraction order will embed the intended phase modulation ϕ. In contrast, if $E_0 \neq 1$ is chosen, there is an additional phase factor found within T_1, namely $\exp(-\mathrm{i}(1 - E_0)\pi)$. This factor can easily be compensated by considering the complex conjugated phase function within the ansatz. Beyond, instead of intended amplitude modulation E_0, the first order reveals an amplitude structure following $\mathrm{sinc}(1 - E_0)$. However, by adjustment of the input amplitude E_0 in Eq. (2.27) by means of a look-up table, the final first diffraction order will contain the desired amplitude as well as phase [73, 75, 80]. In order to spatially separate this first order from the others, an additional blazed grating $\phi_b(x, y)$ (modulus 2π of a linear spatial phase ramp) is included within the phase function. As a result, the first order can be spatially filtered, e.g., by an aperture within the Fourier plane of a $4f$-imaging system. The respective phase hologram is calculated from

$$\Phi_H(x, y) = E_0'(x, y)\,\mathrm{mod}[\phi(x, y) + \phi_b(x, y) - E_0'(x, y)\pi,\, 2\pi] \tag{2.29}$$

with $E_0'(x, y)$ representing the correcting amplitude function. The desired encoded light field is formed in the image plane of the SLM.

Interestingly, blazed gratings do not only allow for the spatial separation of first diffraction order, but also the realization of multiple scalar complex beams by the application of a single hologram [85, 88, 89]. For this purpose multiple blazed gratings $\phi_{b,j}(x, y)$ are combined while each of them carries the information of a chosen spatial mode of amplitude $E_{0,j}(x, y)$ and phase $\phi_j(x, y)$. For this spatial or angular multiplexing approach the respective hologram including corrections is given by

$$\Phi_H^{\text{multi}} = \text{mod}\left[\arg\left(\sum_j^N \mathrm{e}^{\mathrm{i}E'_{0,j}(x,y)[\phi_j(x,y)+\phi_{b,j}(x,y)-E'_{0,j}(x,y)\pi]}\right), 2\pi\right] \tag{2.30}$$

resulting in N modulated light fields.

Note that for best possible modulation results the non-flat surface of the SLM need to be considered. For this purpose a correcting hologram can be determined, e.g., based on an in situ spot correction in the Fourier plane of the SLM plane [80, 90]. Besides improving the scalar complex field realized in the image plane of the SLM, especially the Fourier plane is optimized. This is of special interest if light fields are supposed to be applied in imaging or holographic optical tweezers (HOT) systems, in which the Fourier transformed light field is frequently used[14] [75, 91].

2.4.2 *Interferometric Realization of Vectorial Fields*

Based on the concept of describing vectorial fields as incoherent superposition of two spatial scalar modes (cf. Sect. 2.3.2, Eq. (2.26)), interferometric setups have become established tools for the creation of these beams, in particular for CVBs. For this purpose, one or both of the combined light fields are modulated by a SLM, as described in Sect. 2.4.1, polarized orthogonally and subsequently superimposed on-axis. Various systems have been developed enabling the realization of different vectorial fields [57, 78, 79, 81–85]. An exemplary configuration is sketched in Fig. 2.9a. Here, two beams, beam I and beam II, each of them in its own arm of an interferometer, are shaped by SLMs, SLM_1 and SLM_2, respectively. Thereby, the incident beam is assumed as expanded and collimated. By applying a blazed grating for amplitude and phase modulation (cf. Sect. 2.4.1), the SLMs act as polarization filters based on polarization selective modulation of horizontally aligned liquid crystals (LCs; parallel aligned nematic LC display, e.g., Holoeye Pluto phase-only SLM) [56]—only horizontally polarized light is diffracted in the first order. To polarize beam I and II orthogonally, a half wave plate (HWP) is used after passing SLM_2.

[14]Note that, if light fields, tightly focused by a high numerical aperture (NA ≥ 0.7) optical component, are applied, the focal light field does not represent the Fourier transformed field, since it is formed in the non-paraxial regime. Details on non-paraxial fields are given in Chap. 4.

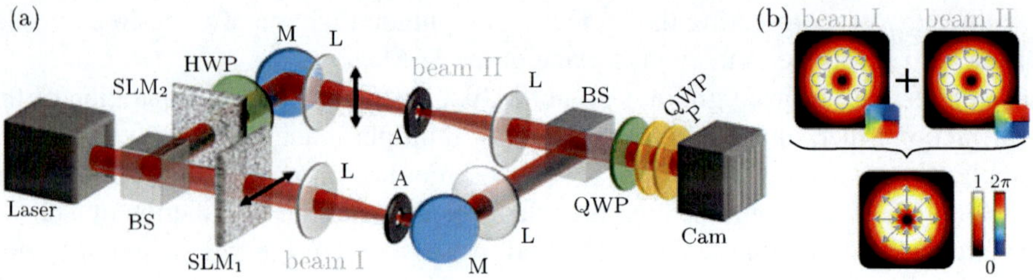

Fig. 2.9 Interferometric approach for the realization of vectorial light. **a** Exemplary sketch of interferometric system with two beams (beam I and II) being shaped by SLMs ($SLM_{1,2}$) and subsequently polarized orthogonally and superimposed on-axis. BS: 50/50 beam splitter, SLM: spatial light modulator, H/QWP: half/quarter wave plate, M: mirror, L: lens, A: aperture, P: polarizer, Cam: camera. Yellow components in combination with camera represent spatially resolved Stokes parameters measurement system. **b** Realization of CVB (radial) by superimposing beam I ($LG^h_{0,1}$, $\vec{e}_R$) and beam II ($LG^h_{0,-1}$, $\vec{e}_L$)

For both beams the first diffraction order is filtered within a $4f$-imaging system built by two lenses (L) and imaging the SLM surfaces on a camera. Beams are joint by a 50/50 beam splitter (BS). Subsequently, on-axis superimposed beams pass through another wave plate, e.g., a quarter wave plate (QWP), so that linear horizontal and vertical polarization are transformed in, e.g., right- and left-circular polarized states. Hence, by for example realizing $LG^h_{0,1}$ and $LG^h_{0,-1}$ as beam I and II, respectively, a radially polarized CVB with first-order (non-generic) V-point is finally formed,[15] as indicated in Fig. 2.9b. In this case the spatially varying phase difference of beam I and II is translated into spatially varying states of polarization.

Obviously, not only LG but various different scalar modes can be generated by $SLM_{1,2}$ so that a broad range of different vectorial singular light fields is accessible. Beyond, applied QWP can be set to different angles resulting in orthogonal states of angle dependent ellipticity or, furthermore, it can be replaced by a wave plate of another chosen retardation. These degrees of freedom further enlarge the range of accessible light fields.

2.4.2.1 Spatially Resolved Polarization Analysis

In order to analyze realized polarization structures in a transverse 2d plane in detail a spatially resolved Stokes parameter measurement system is regularly applied [28, 56, 62, 80]. The system consists of a rotatable QWP and a horizontally aligned polarizer (P) in combination with a detector [92], as indicated in yellow in Fig. 2.9a.

[15]Note that, due to the non-genericity of V-points and experimental constraints of interferometric systems, this singularity is only observed in an idealized case whereas within its experimental formation including perturbations it is probably replaced by two generic C-points as its generic counterparts [7].

For spatially resolved analysis, a camera is used as detector. The intensity detected if the QWP is rotated by β (angle between fast axis and the horizontal) is given by

$$I(\beta) = S_0 + S_1 \cos^2(2\beta) + S_2 \cos(2\beta)\sin(2\beta) + S_3 \sin(2\beta) \tag{2.31}$$

containing normalized Stokes parameters S_{0-3} [92]. Performing a discretization of this equation to intensity values I_v, $v = \{1, 2, 3, ..., N\}$, and applying Fourier analysis results in

$$S_0 = \frac{2}{N}\sum_v I_v - 2I_v \cos(4\beta_v), \qquad S_1 = \frac{8}{N}\sum_v I_v \cos(4\beta_v),$$
$$S_2 = \frac{8}{N}\sum_v I_v \sin(4\beta_v), \qquad S_3 = \frac{4}{N}\sum_v I_v \sin(2\beta_v). \tag{2.32}$$

Hence, by measuring the intensity I_v for at least $N = 8$ different angles β_v, fulfilling sampling theorem, the determination of Stokes parameters is enabled. As the camera detects the intensity spatially resolved, Stokes parameters are determined per camera pixel in the transverse plane of the modulated light field.

2.4.2.2 Simultaneous 2d Customization of Polarization and Energy Flow Density

Besides shaping both, beam I and II in an interferometric system, to be fundamental or complex higher-order Gaussian modes as LG beams, modulation can be restricted to one of the beams (beam I) while the other (beam II) is chosen to be a plane wave [87]. A plane wave is represented by $E_{\text{II}}(\vec{r}) = E_0 \exp(-ikz)$, theoretically being an infinitely expanded wave of homogenoues transverse profile and, thus, even wave front, i.e., $E_0 = \text{const.}$ and its phase $\phi(\vec{r}) = -kz = 0$ for $z = 0$. Obviously, due to experimental constraints, a plane wave experimentally only exists as finite approximation. However, in the following, constant amplitude and transversely homogeneous phase structure of a plane wave for beam II is proposed as a means for the direct observation of amplitude- and phase-to-polarization transfer by an interferometric, thus, superposition approach [87].

As an example, the superposition of horizontally and vertically linearly polarized beams is investigated. Consequently, the QWP in the superimposed beam path of Fig. 2.9a is removed. The initial beam is expanded and collimated. Beam II is vertically polarized ($\vec{e}_{\text{II}} = \vec{e}_y = [0, 1]^T$) and features a plane wave of amplitude $E_0(x, y) = 1 \; \forall x, y$, i.e., $E_{\text{II}}(x, y) = 1$. Hence, SLM_2 is only applied for corrections if demanded. Beam I is horizontally polarized ($\vec{e}_{\text{I}} = \vec{e}_x = [1, 0]^T$) and sculpted by SLM_1 in amplitude and phase with $E_{\text{I}}(x, y) = E_{0,x}(x, y) \cdot \exp(i\phi_x(x, y))$. In the SLMs' image plane, superimposing these fields creates a vectorial beam, tailored in 2d space ((x, y)-plane), with [87]

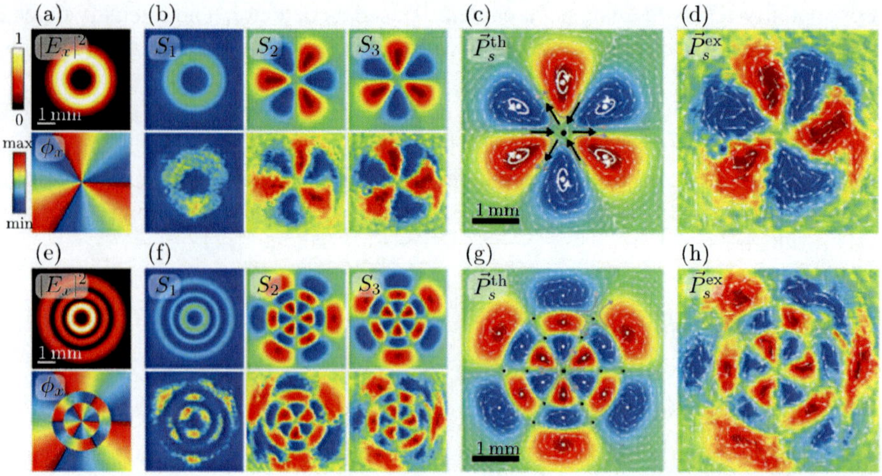

Fig. 2.10 Simultaneous customization of spatial polarization and energy flow in 2d space. **a**, **e** Intensity $|E_x|^2 \in [0,\ 1]$ and phase $\phi_x \in [0,\ 2\pi]$ of beam I ($\mathrm{LG}^h_{0,3}$, $\mathrm{LG}^h_{2,3}$). **b**, **f** Normalized Stokes parameters $S_{1,2,3} \in [-1,\ 1]$ of incoherent superposition of beam I and II. **c**, **g** Theoretically and **d**, **h** experimentally determined SFD $\vec{P}_s^{\mathrm{th,\ ex}}$ (normalized vectors) on S_3. Critical saddle points and centers in elliptical flow fields are marked black and white, corresponding to stationary points of linear polarization and C-point singularities, respectively. [Reproduced with permission from Ref. [87]; © 2018 IOP Publishing. All rights reserved0]

$$\begin{aligned}\vec{E}(x,\ y) = [E_x(x,\ y),\ E_y(x,\ y)]^T &= \vec{e}_{\mathrm{I}}\, E_{\mathrm{I}}(x,\ y) + \vec{e}_{\mathrm{II}}\, E_{\mathrm{II}}(x,\ y)\\ &= \left[E_{0,x}(x,\ y)\cdot \mathrm{e}^{\mathrm{i}\phi_x(x,\ y)},\ 1\right]^T. \end{aligned} \tag{2.33}$$

Obviously, polarization varies spatially, directly dependent on amplitude $E_{0,x}$ and phase ϕ_x. Two illustrative examples are given in Fig. 2.10a, b and e, f with $E_{\mathrm{I}} = \mathrm{LG}^h_{0,3}$ and $E_{\mathrm{I}} = \mathrm{LG}^h_{2,3}$, respectively.[16] Here, Fig. 2.10a and e show the according intensity and phase distributions of beam I, Fig. 2.10b and f the respective theoretically calculated (top, Eq. (2.22)) and experimentally determined (bottom) Stokes parameters S_{1-3}. The latter were determined following above explained measurement procedure. The experimental results reflect theoretical calculations, highlighting the inhomogeneous polarization structure. While S_1 clearly mirrors the intensity distribution of beam I, S_2 and S_3 reveal a petal structure[17] surrounding the beam axis due to the combination of $|E_x|^2$ and ϕ_x interacting with E_{II}. Note that petals are placed on n rings with $2|\ell|$ petals per ring, evincing the dependence on the mode indices n and ℓ of the LG mode in beam I. Hence, by adapting these indices, the resulting

[16] Note that within the experiment a reflective SLM was applied as SLM_1 for LG mode generation, whereas SLM_2 was removed.

[17] Deviations in petal orientation in experiment compared to theory originates from phase shifts between beam I and II due to sensitivity of interferometric systems (temporal instability).

polarization configuration can be tailored on demand confirming an amplitude- and phase-to-polarization relation [87].

In addition, strongly connected to the amplitude- and phase-to-polarization relation, another feature can be observed by the above explained structures: the translation of so-called orbital flow density (OFD) to spin flow density (SFD) [87]. As outlined in Sect. 2.2.3, helical LG modes carry OAM depending on mode index ℓ. The optical angular momentum is the mechanical attribute of the transverse energy flow (TEF) embedded in these structures. Besides affecting, e.g., particles trapped by light, TEF is a meaningful and informative characteristic of light fields, since it is related to the fields' spatial structure including its optical singularities. As a consequence, it can be used as an illustrative and easily interpretable, instructive representation of the singular field properties [93–95].

Generally, the direction and density of energy transport in light (energy flow density/linear momentum density $\vec{p}$) is determined by the Poynting vector $\vec{S}_P = \vec{E} \times \vec{H}$ with $\vec{p} = \vec{S}_P/c^2$ [96, 97]. Thereby, the total TEF density $\vec{P}$ is composed of its orbital and spin parts, i.e., the OFD ($\vec{P}_o$) and SFD ($\vec{P}_s$). Under paraxial approximation these are represented by [96]

$$\vec{P}_o = \frac{1}{k}(I_x \nabla\phi_x + I_y \nabla\phi_y), \qquad \vec{P}_s = -\frac{1}{2k}(\vec{e}_z \times \nabla S_3), \tag{2.34}$$

with $\nabla = [\partial/\partial x, \partial/\partial z, \partial/\partial z]^T$, intensity $I_{x,y} = |E_{x,y}|^2$ and the longitudinal unit vector $\vec{e}_z = [0,\ 0,\ 1]^T$. This equation clearly reveals the relation of OFD and SFD to the phase and polarization of considered light field, respectively. Clearly, applied LG modes embedding helical phase structures feature azimuthally pointing OFD. This OFD is partially translated into SFD by the superposition approach [87], since the spatial amplitude and phase structure is transformed into spatially varying polarization. The inital beams I and II include

$$\text{beam I:} \qquad \vec{P}_o = \frac{1}{k}(I_x \nabla\phi_x), \qquad \vec{P}_s = \vec{0} \tag{2.35}$$

$$\text{beam II:} \qquad \vec{P}_o = \vec{0}, \qquad \vec{P}_s = \vec{0}. \tag{2.36}$$

Following Eq. (2.33), the OFD and SFD of the field shaped by their superposition (beam I + II) is given by [87]

$$\text{beam I + II:} \qquad \vec{P}_o = \frac{1}{k}(I_x \nabla\phi_x), \qquad \vec{P}_s = -\frac{1}{2k}(\vec{e}_z \times \nabla S_3). \tag{2.37}$$

Thus, by combining two beams of no SFD, but one of them including an OFD structure, a light field is created which carries SFD due to $S_3 \neq 0$. In particular, a transversely inhomogeneous S_3 or spin configuration with [87]

$$S_3(x, y) \propto -2E_{0,x}(x, y)\sin(\phi_x(x, y)) \tag{2.38}$$

is formed. According to Eq. (2.34) the respective SFD follows the shape of S_3 as demonstrated in Fig. 2.10c, d and g, h (theoretically, experimentally) for utilized $\mathrm{LG}^h_{0,3}$ and $\mathrm{LG}^h_{2,3}$, respectively. Here, the normalized spin flow $\vec{P}_s$ is illustrated by white arrows on S_3. Within the conservative flow system of closed trajectories [32] critical points can be observed. On the one hand, unstable saddle[18] or hyperbolic points [6, 95, 96], marked black, can be found on-axis and located on rings around the beam axis. Note that on-axis saddles are of higher-order as instead of typically observed two stable and two unstable radially oriented trajectories of a standard saddle point, there are $|\ell|$ of each stability class. On the other hand, as characteristic for conservative systems, "centers" (also called "elliptic points", "circulations" or "vortices" [6, 95, 96, 98]; marked white) of elliptical flow fields are positioned in global extrema of S_3 ($= \pm 1$). Consequently, centers in SFD reveal C-point singularities in these vectorial light fields. Beyond, observed saddle points correspond to stationary points of linear polarization within the polarization configuration. Hence, also in this case the energy flow can be utilized as an easily interpretable and visual tool for the identification of vectorial singularities in structured light. Depending on characteristics of beam I, the OFD to SFD translation can be adjusted, so that different configurations of critical points and, thus, polarization singularities are realized [87].

Note that, similar to so-called Poynting vector singularities [99], critical points observed within configuration above can also be described as generic or non-generic optical singularities—in this case, vectorial SFD singularities. Under paraxial approximation, the SFD is defined as $\vec{P}_s = [P_{s,x}, P_{s,y}, 0]^T = [\partial S_3/\partial y, -\partial S_3/\partial x, 0]^T/2k$ (see Eq. (2.34)). Hence, it represents a 2d vector in the 2d transverse plane. Here, for the occurrence of a SFD singularity, the vector $\vec{P}_s$ and, thus, its components $P_{s,x}$ and $P_{s,y}$ need to vanish ($= 0$). As a consequence, a complex field according to Eq. (2.1) can be defined by $\Gamma_s = P_{s,x} + \mathrm{i}P_{s,y}$ with a singular point corresponding to a $\Phi_s = \arg(\Gamma_s)$ phase singularity with index $n_s = \oint_C \mathrm{d}\Phi_s/2\pi$. A singularity occurs for intersections of zero-lines $Z_{\Im}$ ($P_{s,y} = 0$) and $Z_{\Re}$ ($P_{s,x} = 0$). Hence, the codimension of the SFD singularity is two, i.e., its generic form is a point in 2d space with $n_s = \pm 1$. For $n_s = 1$ a critical point representing a center (right-/left-handed) is observed. In contrast, $n_s = -1$ corresponds to standard saddle points.[19] Importantly, higher-order saddle points with $|n_s| > 1$, as observed on-axis for structures in Fig. 2.10, represent non-generic SFD singularities since for their occurrence multiple zero-lines need to cross. Thus, these higher-order saddle points will unfold into first-order generic ones under perturbation, e.g., within the presented experiment including free space propagation. As an illustrative example, the zero-lines on S_3 and the phase Φ_s are shown in Fig. 2.11a, b, respectively, for the vectorial field created by superposition of a $\mathrm{LG}^h_{2,3}$ mode with a plane wave (numerics; cf. Fig. 2.10e–h). Clearly, zero-lines cross multiple times on-axis, corresponding

[18]Critical saddles in energy flow represent singularities in these flow systems and are not to be confused with saddle points of polarization in vectorial light fields (cf., e.g., Sect. 3.1).

[19]Centers and saddles correspond to a conservative system. In contrast, if absorption within the system is considered, critical points with $n_s = 1$ may represent an inward node, a sink, or a (left-/right-handed) spiral [99].

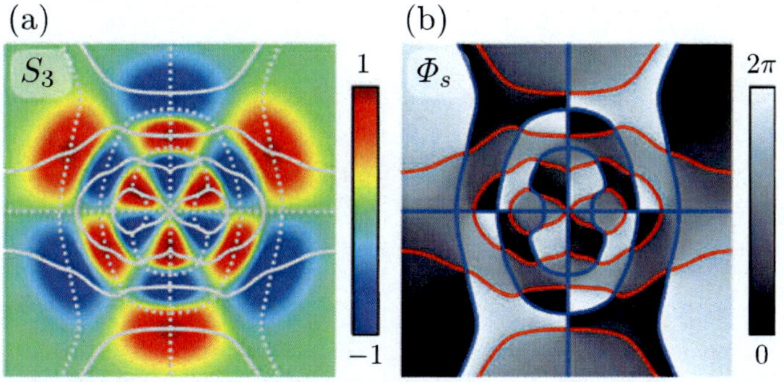

Fig. 2.11 Detection and analysis of SFD singularities. Zero-lines are shown on **a** S_3 and **b** $\Phi_s = \arg(\Gamma_s)$ ($Z_{\mathfrak{I}}$: dashed/blue, $Z_{\mathfrak{R}}$: continuous/red) revealing generic SFD singularities with $n_s = \pm 1$ (saddle or center) and non-generic ones with $n_s = 2$ (higher-order on-axis saddle)

to a non-generic saddle point with $n_s = 2$, whereas off-axis zero-line intersections reveal standard generic centers and saddles.

2.4.2.3 Conclusion

An established method based on the superposition principle was presented for the creation of vectorial light fields. Its functionality was proven by shaping vectorial fields in 2d space by incoherently combining a plane wave and higher-order LG modes. Beyond, not yet known spin flow density (SFD) configurations were tailored by translation of OAM structures to spatially varying SAM, thus, of orbital to spin flow density [87]. Shaped structures embed an on-demand network of critical points, namely generic and non-generic SFD singularities. By adapting LG mode indices the number, position and order of critical saddle and elliptical points (centers) in the conservative flow system and, hence, polarization singularities can be customized. Besides being an illustrative tool for the identification of polarization singularities, this approach is of particular interest for advanced optical trapping schemes as trapped particles could be guided on-demand by respective energy flow structures.

Even though interferometric approaches allow for a variety of different vectorial fields including singularities and even for energy flow customization, a key challenge represents the exact on-axis superposition of scalar modes reasoning the strong sensitivity to errors [7]. This sensitivity impedes the study of singularities, especially non-generic ones, as they do not occur in an error-prone system of misaligned superimposed beams. To overcome this shortcoming, different techniques have been developed (e.g.., see Refs. [54, 56, 57, 76, 78, 80, 100]) ensuring a more precise superposition or even allowing for a single-beam approach for polarization modulation. In the following, a dynamic, single-beam system is outlined facilitating the customization of intriguing polarization structures and singularities.

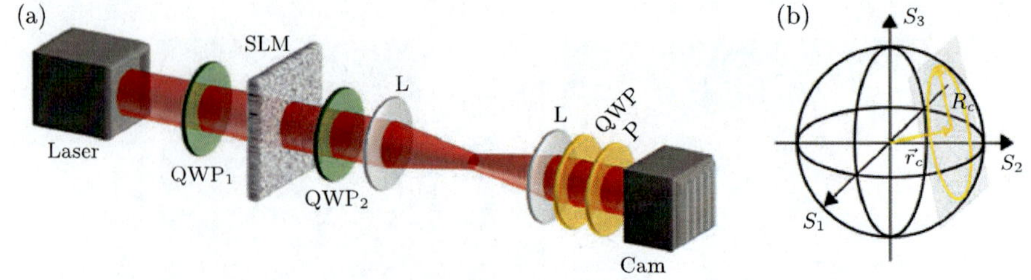

Fig. 2.12 Dynamic polarization modulation system (DPMS). **a** Sketch of experimental system including measurement setup (QWP: quarter wave plate, SLM: spatial light modulator, L: lens, P: polarizer, Cam: camera, yellow: components of measurement system). Experimentally, a reflective SLM is applied. **b** Subspace of accessible polarization states on Poincaré sphere

2.4.3 Dynamic Polarization Modulation

Besides structuring two individual beams and, subsequently, incoherently superimpose them, one can also choose to spatially shape the orthogonal polarization components of a single beam separately without the need of spatially splitting them. For this purpose the dynamic polarization modulation system (DPMS) can be applied [56, 80], which consists of a phase-only SLM (horizontally parallel aligned nematic LC display) as key component enclosed by two wave plates, e.g., QWPs.[20] A sketch of this system is shown in Fig. 2.12a. Considering initial horizontally linear polarization of an expanded collimated beam, the first QWP_1 and its orientation β_1 define the ratio of horizontal and vertical polarization components being sent to the SLM. The orientation β_1 is defined as the angle (counterclockwise) between the input polarization and the fast axis of the wave plate. Subsequently, as the SLM only modulates horizontal components, whereas vertical ones stay unaffected, the SLM introduces a phase shift $\Delta\Phi_{\text{SLM}} \in [0, 2\pi]$ between these orthogonal parts, i.e., the polarization is changed.[21] In a final step, the second QWP_2 is used to recombine horizontal and vertical components dependent on its orientation β_2. All states of polarization accessible in the beam's transverse 2d plane are located on a circular ring of radius R_c and central position $\vec{r}_c$ on the Poincaré sphere, as indicated in Fig. 2.12b. By the adaption of the orientation angle $\beta_{1,2}$ of $\text{QWP}_{1,2}$ the radius $R_c(\beta_1)$ and position $\vec{r}_c(\beta_2)$ of the ring can be adjusted [56, 80]. Here, each point on the ring corresponds to another phase shift $\Delta\Phi_{\text{SLM}}$, consequently, by applying a spatially varying phase hologram $\Delta\Phi_{\text{SLM}}(x, y)$ light fields of inhomogeneous polarization are tailored in 2d space.

---

[20]Note that wave plates of different retardation can be incorporated [56]. However, within this thesis, solely the configuration with QWPs is used. Further, a reflective SLM is applied in contrast to the simplified sketch in Fig. 2.12a.

[21]No blazed grating is used here.

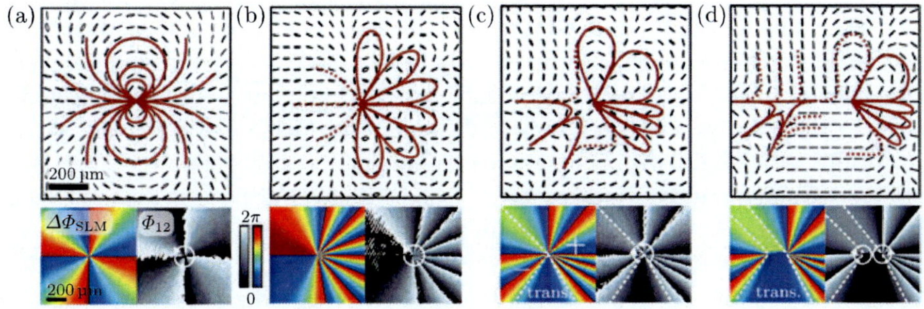

Fig. 2.13 Vector beams tailored by the DPMS with hologram $\Delta\Phi_{\text{SLM}}$ being directly related to phase Φ_{12}. **a–c** Experimentally realized polarization structures (top; red: flow lines) are determined by Stokes parameter measurement. **a** Two-petal flower ("spider") structure with $\sigma_{12} = 4$, **b** irregular six-petal flower configuration with $\sigma_{12} = 8$, and **c** hybrid structure with two very close singularities of index $\sigma_{12} = 6$ and -2, thus, $\sigma_{12}^{\text{tot}} = 4$ (dashed line: boarderline of positive and negative phase gradient; trans.: transition zone). **d** Simulation (Jones calculus) of vector field in **c** with increased distance of singularities

2.4.3.1 Tailored Vector Fields in 2d Space

For the realization of vector fields by the DPMS the ring-shaped subspace have to be positioned on the equator of the Poincaré sphere. For this purpose, angles $\beta_{1,2}$ are set to $\pm\pi/4$ ($\pm 45°$) [56, 62]. Interestingly, if for both, QWP_1 and QWP_2, angles are chosen to be negative with $\beta_{1,2} = -\pi/4\,(-45°)$, the hologram of the reflective SLM is directly related to the resulting light field with $\Delta\Phi_{\text{SLM}} = \Phi_{12}$. This relation facilitates a straightfoward realization of arbitrary vector fields with singularities whose number, position and index can be tailored on demand [62]. Figure 2.13 depicts some (a)–(c) experimental and (d) numerical examples with applied holograms[22] $\Delta\Phi_{\text{SLM}}$ and resulting phase Φ_{12} (bottom) presented below respective polarization distributions with red flow lines (top). Experimentally the polarization distribution is determined by the Stokes parameter measurement system (camera and yellow components in Fig. 2.12a) with the camera being placed in the image plane of the SLM ($z = 0$).

Obviously, for all experimental results (a)–(c) the measured phase Φ_{12} perfectly mirrors the respective holograms $\Delta\Phi_{\text{SLM}}$, revealing the high spatial resolution and quality of DPMS realized fields. As a consequence, the polarization structures reveal a very clear shape. Here, Fig. 2.13a shows a regularly shaped two-petal flower structure also called "spider" with a central V-point singularity of index $\sigma_{12} = 4$. Note that non-generic singularities in fields of Fig. 2.13a–c are only formed in the exact plane of the SLM (undisturbed). Upon propagation these singularities unfold (cf. Chap. 3), whereby the singularity index σ_{12} of the field is kept. Since the SLM allows for arbitrary holograms ($\Delta\Phi_{\text{SLM}} \in [0, 2\pi]$), the deformation of vector fields is enabled as demonstrated in Fig. 2.13b. By the choice of a higher-order phase vortex with a non-linear azimuthal phase gradient as hologram, a deformed irregular flower structure is

[22]Note that for optimal results an additional correction hologram is applied here as explained in Ref. [80].

formed, embedding a V-point with $\sigma_{12} = 8$. Both, regular (a) as well as irregular (b) vector fields follow the petal-to-index relation, i.e., $|\sigma_{12} - 2|$ petals are obsevered for both [62]. The same rule is applicable for web configuration. As indicated in Sect. 2.3.2, this rule is not kept for hybrid structures, which is experimentally proven by results in Fig. 2.13c with a hologram merging two vortices of opposite handedness. Here, two singularities of indices $\sigma_{12} = 6$ and -2 are embedded in the joint vector field, hence, $\sigma_{12}^{\text{tot}} = 4$. However, according to indices, there are $|\sigma_{12} - 2| = 4$ petals and sectors expected, but five and two are observed, respectively, due to the too close vicinity of V-points [7, 62]. If their distance is increased, as visualized in Fig. 2.13d by numerical results determined by Jones calculus (details see following paragraph), the petal-to-index relation is recovered.

2.4.3.2 Simultaneous System Dependent Phase Modulation

As demonstrated above, the DPMS allows for on-demand realization of tailored singular vector fields. Of course by adaption of angles $\beta_{1,2}$ various polarization structured fields—ellipse as well as vector fields—with chosen singularities can be shaped [28, 56, 80]. Note that besides intended polarization customization the DPMS simultaneously modulates the phase of the respective field [28]. This system dependent phase modulation $\phi_{\text{DPMS}}(x, y)$ can be determined easily by Jones calculus. As an illustrative example, $\phi_{\text{DPMS}}(x, y)$ is computed for $\beta_{1,2} = -\pi/4$, i.e., if vector fields are created as shown above. The SLM and $\text{QWP}_{1,2}$ ($\beta = \beta_{1,2} = -\pi/4$) are represented by Jones matrices

$$J_{\text{SLM}}(\Delta\Phi_{\text{SLM}}(x, y)) = \begin{pmatrix} -\,e^{-i\Delta\Phi_{\text{SLM}}(x,y)} & 0 \\ 0 & 1 \end{pmatrix}, \tag{2.39}$$

$$J_{\text{QWP}}(\beta) = e^{\frac{-i\pi}{4}} \begin{pmatrix} \cos^2\beta + i\sin^2\beta & (1-i)\sin\beta\cos\beta \\ (1-i)\sin\beta\cos\beta & \sin^2\beta + i\cos^2\beta \end{pmatrix} = \frac{1}{\sqrt{2}}\begin{pmatrix} 1 & i \\ i & 1 \end{pmatrix}. \tag{2.40}$$

Assuming the input light field $\vec{E}^{\text{in}}$ to be a plane wave (normalized amplitude $E^{\text{in}}_{x,y,0}(x, y) = 1$, even phase front $\phi^{\text{in}}_{x,y}(x, y) = 0$) of horizontal linear polarization, i.e., $\vec{E}^{\text{in}} = [1, 0]^{\text{T}}$, matrix calculations result in the modulated electric field [28]

$$\begin{aligned} \vec{E}(x, y) &= J_{\text{QWP}}(\beta_1 = -\pi/4) \cdot J_{\text{SLM}}(\Delta\Phi_{\text{SLM}}) \cdot J_{\text{QWP}}(\beta_2 = -\pi/4) \cdot \vec{E}^{\text{in}} \\ &= \frac{1}{2}\begin{pmatrix} -e^{-i\Delta\Phi_{\text{SLM}}(x,y)} - 1 \\ -i\,e^{-i\Delta\Phi_{\text{SLM}}(x,y)} + i \end{pmatrix}. \end{aligned} \tag{2.41}$$

Hence, in addition to intended polarization modulation in $\vec{E}$ the x- (horizontal) and y-component (vertical) of the electric field have a joint phase of

$$\phi_{\text{DPMS}} = -\Delta\Phi_{\text{SLM}}/2 \tag{2.42}$$

with

$$\phi_x = -\Delta\Phi_{\text{SLM}}/2 + \begin{cases} \pi & \text{for } \Delta\Phi_{\text{SLM}} \in [0, \pi]\,, \\ 0 & \text{for } \Delta\Phi_{\text{SLM}} \in]\pi, 2\pi]\,, \end{cases}$$
$$\phi_y = -\Delta\Phi_{\text{SLM}}/2 + \pi \quad \text{for } \Delta\Phi_{\text{SLM}} \in [0, 2\pi[\,. \tag{2.43}$$

2.4.3.3 Conclusion

Besides outlining the concept of the DPMS, innovative vector fields were presented experimentally, revealing new insights into the fundamentals of singular optics. In contrast to typically studied higher-order vector fields of regular axial symmetric shape, irregular and even hybrid configurations were formed [62]. It has been proven that, equivalent to regular structures, irregular flower- or web-shaped vector fields reveal $|\sigma_{12} - 2|$ petals or sectors. This observation confirms the relation of the light fields structure and the index of embedded singularity. In contrast, for hybrid fields, imparting a negatively and positively indexed V-point singularity in very close vicinity, experimental results manifest that the rule of $|\sigma_{12} - 2|$ petals and sectors is not kept. However, it was demonstrated that if the distance of singularities is increased, the rule is recovered.

Additionally, the simultaneous system dependent phase variation by the DPMS was outlined mathematically. This additional feature of the DPMS can be applied for, e.g., studies on combined phase and polarization structuring (see Sect. 3.2) [28]. If purely polarization structured fields are desired, different methods exist in order to compensate for the additional phase modulation. For example the DPMS can be combined with a phase or even amplitude and phase modulation system [56, 80], as applied in, e.g., Sect. 3.3.2. Here, the SLM with high-definition (HD) display (1920×1080 px) operates in split-screen mode with the beam passing the SLM twice. In the first pass amplitude and phase is shaped by the first SLM half (cf. Sect. 2.4.1) and in the second pass the second SLM half in combination with two QWPs constitutes a DPMS. Hence, if desired, ϕ_{DPMS} can be compensated in advance within the first pass. Beyond, this configuration can be used to tailor amplitude, phase and polarization ("APP" modulation system), i.e., to create fully-structured light fields. In the following another sophisticated method is proposed, which has the advantage of operating in full-screen mode, thus, modulation is performed with optimized spatial resolution.

2.4.4 *Multiplexing Approach for Sculpting Light*

Here, an advanced holographic single-beam method is proposed which does not only facilitate simultaneous direct modulation of amplitude, phase and polarization but also ensures high spatial resolution using standard, thus affordable, non-HD

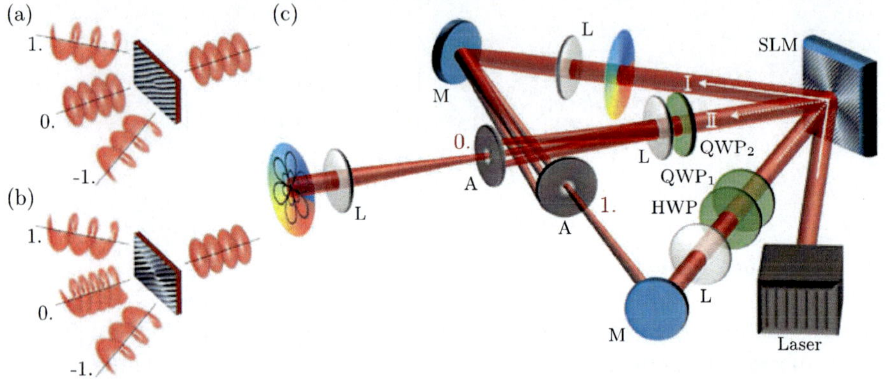

Fig. 2.14 Spatial multiplexing approach for combined phase and polarization modulation. Sketch of encoding phase structures **a** in the first diffraction order by a blazed grating, or **b** in the first and zeroth diffraction order by a multiplexed hologram according to Eq. (2.44). **c** Schematics of experimental system applying the multiplexing approach for combined modulation of phase and polarization (I/II: first/second SLM pass, SLM: spatial light modulator, M: mirror, L: lens, A: aperture, H/QWP: half/quarter wave plate). [Reproduced with permission from Ref. [86]; © 2018 IOP Publishing. All rights reserved]

SLMs in full-screen-mode [86]. This method is based on spatial multiplexing of different holograms onto the SLM (cf. Sect. 2.4.1), which is passed twice. In contrast to split-screen techniques, the beam hits the SLM (here: Holoeye Pluto phase-only SLM) at the same position in both passes. The first pass (I) is used for amplitude and phase modulation encoded in the first diffraction order (cf. Sect. 2.4.1). The second pass (II) modulates polarization by a DPMS utilizing the zeroth diffraction order (cf. Sect. 2.4.3). Hence, the hologram needs to carry the information for both passes, but sent the according information into different orders.

First, the combined modulation of phase and polarization is considered. For this purpose, the ansatz of spatial multiplexing by blazed gratings as information carriers is used as basis, as presented in Sect. 2.4.1, Eq. (2.30). In contrast to the typical approach, only the intended phase $\phi(x, y)$ modulation is saved in a blaze grating $\phi_b(x, y)$, whereas the phase information $\Delta\Phi_{\mathrm{SLM}}(x, y)$ needed for polarization structuring is encoded without blazed grating. Thus, following Eq. (2.30) the multiplexed hologram is represented by [86]

$$\Phi_H(x, y) = \mathrm{mod}\left[\arg\left(\mathrm{e}^{\mathrm{i}[\phi(x,y)+\phi_b(x,y)]} + \mathrm{e}^{\mathrm{i}\Delta\Phi_{\mathrm{SLM}}(x,y)}\right),\ 2\pi\right]. \tag{2.44}$$

Typically, if a blazed grating is applied for pure phase modulation, there is no modulation observed within the zeroth order but in the first (cf. Eq. (2.28)), as indicated in Fig. 2.14a.[23] Hence, if the multiplexed hologram in Eq. (2.44) is used, the zeroth order solely includes the intended modulation $\Delta\Phi_{\mathrm{SLM}}$, while the first embeds ϕ, as

[23] A simplified sketch is shown. For the experimentally applied (asymmetric) blazed grating the intensity of $+1.$ and $-1.$ order differs.

visualized in Fig. 2.14b. The respective setup of combined phase and polarization modulation is depicted in Fig. 2.14c with the multiplexed hologram being displayed on the SLM. The horizontally polarized initial laser beam (expanded and collimated) hits the SLM the first time, being modulated in phase (pass I). Subsequently, by a $4f$-imaging system of two lenses (L), in which the first diffraction order is filtered by an aperture (A) in Fourier plane, the SLM is imaged onto itself. This imaging system initiates the second pass (II) in which the SLM is enclosed by QWP_1 and QWP_2 for constituting a DPMS. Hence, the phase modulated beam is shaped by the DPMS resulting in a phase and polarization structured light field. This field is observed in the zeroth order of the second pass, again filtered by an aperture within a $4f$-system. Note that in pass II parts of the horizontally polarized light is diffracted in higher orders due to the multiplexed hologram, whereas vertically polarized components fully stay in the zeroth order. To compensate for losses of horizontal components, a HWP is applied in front of QWP_1 correcting the ratio of vertical and horizontal parts by adjusting the angles of the wave plates set [86].

2.4.4.1 Functionality of Multiplexed Hologram—Double-Phase Modulation

For demonstrating the functionality of respective multiplexed holograms, the introduced experimental system is applied for "double-phase modulation", i.e., both passes are used to shape the phase. For this purpose, the wave plates are removed from the experimental setup so that a phase modulation of $\phi(x, y) + \Delta\Phi_{\text{SLM}}(x, y)$ is expected. The resulting field is observed in the image plane of the SLM by a camera. Here, the created phase modulation is determined by interference with a reference beam (plane wave, horizontal polarization) under a small angle [101]. As illustrative examples the following settings are chosen

$$\begin{aligned}
&1) \quad \phi(x, y) = \ell\varphi, \quad && \Delta\Phi_{\text{SLM}}(x, y) = 0, \\
&2) \quad \phi(x, y) = 0, \quad && \Delta\Phi_{\text{SLM}}(x, y) = \ell\varphi, \\
&3) \quad \phi(x, y) = \ell\varphi, \quad && \Delta\Phi_{\text{SLM}}(x, y) = \ell\varphi.
\end{aligned}$$

with $\ell = 2$ and the polar angle φ. The respective experimental (top) and numerical (bottom right) results are presented in Fig. 2.15a with (a1)–(a3) corresponding to setting 1)–3). Numerical results are calculated from the respective hologram Φ_H (cf. Eq. (2.44); diffraction efficiency ~100%). Obviously, for all settings the results prove the expected phase modulation of $\phi(x, y) + \Delta\Phi_{\text{SLM}}(x, y)$. All vortex configurations perfectly reveal an azimuthally linear phase gradient—in theory and experiment. Hence, the effective functionality of the multiplexed hologram Φ_H is evinced.

In contrast, if a SLM of less diffraction efficiency is applied (e.g., Liquid Crystal on Silicon Modulator of Boulder Nonlinear Systems), results show deviations from expected structures [86], as proven in Fig. 2.15b. Here, settings 1) and 2) are applied corresponding to Fig. 2.15b1 and b2, respectively. While for setting 1) the phase

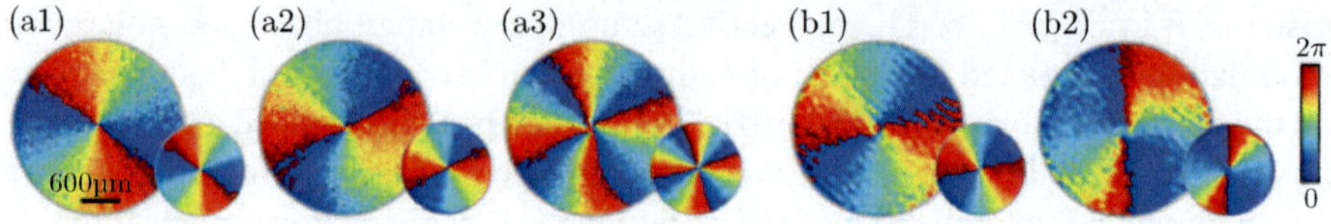

Fig. 2.15 Functionality of multiplexed hologram visualized by double-phase modulation. **a** Experimental (top; Holoeye Pluto SLM) and numerical (bottom right) results of relative phase measurement for double phase modulated light field with (a1)–(a3) corresponding to setting 1)–3). **b** Phase measurement for lower diffraction efficiency (60%, Liquid Crystal on Silicon Modulator of Boulder Nonlinear Systems) and settings 1) and 2)

vortex still shows an azimuthally linear change in phase, for 2) a non-linear phase gradient is observed. This effect is due to the fact that for lower diffraction efficiencies, a blazed grating as used for encoding ϕ does not diffract 100% into the first diffraction order, but there remains light in the zeroth. Consequently, the modulation observed in the zeroth order corresponds to the superposition of a plane wave of amplitude $E_0^\star$ and the intended modulated field, i.e., $E_0^\star + \exp(\mathrm{i}\Delta\Phi_{\mathrm{SLM}}(x, y))$. Result in Fig. 2.15b match a diffraction efficiency of 60%.

2.4.4.2 Customized Phase and Polarization Modulation

To demonstrate the ability to customize phase and polarization by the multiplexing approach, the realization of DPMS modulated polarization structures with compensated system dependent phase variation ϕ_{DPMS} (cf. Sect. 2.4.3, Eq. (2.42)) is shown [86]. For this purpose, a lowest- as well as higher-order vector field, namely a radially polarized field as well as a regularly shaped flower configuration, is investigated. Hence, wave plates with $\beta_{1,2} = -\pi/4$ are placed into the experimental system and $\phi(x, y) = 0$ is chosen first, so that solely polarization customization by the DPMS is observed. As shown in Fig. 2.16a, d, if $\Delta\Phi_{\mathrm{SLM}} = \sigma_{12}\varphi$ is chosen, modulation results in a radial beam for $\sigma_{12} = 2$ and a flower structure for $\sigma_{12} = 8$. Polarization distributions (left; red: flow lines) determined by Stokes parameter measurement as well as respective complex Stokes field's phase $\Phi_{12} = \Delta\Phi_{\mathrm{SLM}}$ (right) are presented. The point symmetric, regular structure of polarization states and the linear phase gradient in Φ_{12} evince the quality of polarization modulation. As mathematically demonstrated in Sect. 2.4.3, the DPMS simultaneously create an additionally system dependent phase variation $\phi_{\mathrm{DPMS}} = -\Delta\Phi_{\mathrm{SLM}}/2$. Hence, for the radial or flower-shaped beam an additional phase vortex of topological charge $\ell = -\sigma_{12}/2 = -1$ or -4, respectively, is imprinted. This can be shown by near-field single slit diffraction (slit width: 0.06 mm), as illustrated in Fig. 2.16b, e for the radial and flower-shaped beam, respectively. The slit is positioned in the image plane of the SLM (white dashed line in Fig. 2.16a) with the camera now being placed slightly behind this plane (about 3 mm). Here, the embedded helical phase front of ϕ_{DPMS} causes the near-field image of the slit to be dislocated according to the topological charge ℓ of the vortex, as

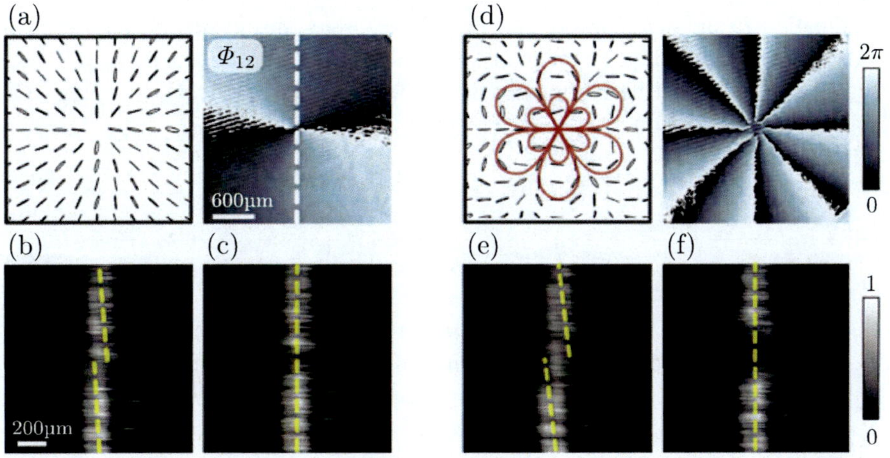

Fig. 2.16 Simultaneous modulation of phase and polarization by the multiplexing approach. **a**, **d** Measured polarization distribution and respective phase Φ_{12} of generated radial and flower-shaped vector field. Near-field slit diffraction shows **b/e** dislocated or **c/f** clear image of slit for non-compensated or compensated system dependent phase variation ϕ_{SLM}, respectively. Position of slit indicated by dashed white line in **a**. [Reproduced with permission from Ref. [86];]

emphasized by the yellow dashed line. To compensate for this phase modulation, the multiplexing approach is performed. By additional customization of phase with $\phi(x, y) = \frac{\sigma_{12}}{2}\varphi$ within the first SLM pass, the system dependent phase variation is removed, as proven by slit diffraction in Fig. 2.16c and f.

2.4.4.3 Natural and On-Demand Amplitude Structuring

Light fields are referred to as fully-structured if they include a spatially inhomogeneous distribution in amplitude, phase and polarization. Above demonstrated fields are tailored in phase and polarization, whereby they can additionally reveal natural or on-demand amplitude structuring. The former one is realized easily by free space propagation of the field customized in phase and polarization. This is for example observable in Fig. 2.16f. Here, the intensity configuration reveals a dark area on axis, which is formed by unfolding higher-order, thus, non-generic singularities (more details in Chap. 3) in the tailored light field. Consequently, the amplitude is naturally shaped by free space propagation.

Beyond, the introduced multiplexing approach can be upgraded to additionally incorporate direct on-demand amplitude sculpting [86]. For this purpose, the intended amplitude information $E_0(x, y) \in [0, 1]$ is additionally encoded in the blazed grating of the first pass following the approach presented in Sect. 2.4.1. Hence, first amplitude and phase is modulated, second the polarization is structured in pass II. However, the amplitude encoding in the blaze grating causes an approximately inverse amplitude modulation in the zeroth order (cf. Eq. (2.28)), thus, in the second pass used for polarization modulation. To make of for this undesired amplitude structuring within pass II, an additional beam function is required within Eq. (2.44), resulting in the new hologram function

$$\Phi_H(x,y) = \mathrm{mod}\left[\arg\left(\mathrm{e}^{\mathrm{i}E_0'(x,y)[\phi(x,y)+\phi_{b,1}(x,y)-E_0'(x,y)\pi]} + \mathrm{e}^{\mathrm{i}\Delta\Phi_{\mathrm{SLM}}(x,y)} \right.\right.$$
$$\left.\left. + \mathrm{e}^{\mathrm{i}E_{0,\mathrm{corr}}'(x,y)[\phi_{b,2}(x,y)-E_{0,\mathrm{corr}}'(x,y)\pi]}\right),\ 2\pi\right]. \quad (2.45)$$

Here, E_0' and $E_{0,\mathrm{corr}}'$ represent the amplitude and correcting amplitude following a look-up table (see Sect. 2.4.1), encoded in blazed gratings $\phi_{b,1}$ and $\phi_{b,2}$, respectively. By this ansatz a non-interferometric single-beam technique is proposed for the on-demand realization of fully-structured light with high spatial resolution.

2.4.4.4 Superposition Principle for Vector Modes

Besides representing an innovative tool for the customization of fully-structured light with high spatial resolution, the presented system can also be applied for the realization of vector fields following the superposition principle. For this purpose the first QWP_1 is removed from the system presented in Fig. 2.14c and the HWP is set to $\beta = \pi/8$ (22.5°). The used multiplexed hologram follows Eq. (2.44). As representative example the intended light field is assumed to be a radially polarized CVB (cf. Eq. (2.26)), defined as

$$\vec{E}_{\mathrm{CVB}} = \frac{1}{\sqrt{2}}\left[\vec{e}_R \cdot \mathrm{LG}^h_{0,1} + \vec{e}_L \cdot \mathrm{LG}^h_{0,-1}\right]. \quad (2.46)$$

Here the incident light field is considered as a Gaussian mode $E_G = \mathrm{LG}^h_{0,0}$ (cf. Eq. (2.11)). First, in pass I a helical phase front with $\ell = -1$ is encoded on this horizontally polarized beam ($\phi(x,y) = -\varphi$ in Eq. (2.44)). Subsequently, after filtering the first order, the HWP is used to rotate the polarization diagonally. Hence, the respective light field is now represented by

$$\vec{E}_{\mathrm{CVB},1} = \frac{1}{\sqrt{2}}\left[\vec{e}_H \cdot \mathrm{LG}^h_{0,0}\,\mathrm{e}^{-\mathrm{i}\varphi} + \vec{e}_V \cdot \mathrm{LG}^h_{0,0}\,\mathrm{e}^{-\mathrm{i}\varphi}\right].$$

with $\vec{e}_H = [1,\ 0]^T$ and $\vec{e}_V = [0,\ 1]^T$. In the next step, this field or, more precisely, its horizontally polarized part is modulated by the SLM imprinting an additional phase vortex of topological charge $\ell = +2$ ($\Delta\Phi_{\mathrm{SLM}}(x,y) = 2\varphi$ in Eq. (2.44)). The vertical part is solely reflected propagating along the same pass as horizontal components (zeroth diffraction order). Thus, $\vec{E}_{\mathrm{CVB},1}$ is transformed into

$$\vec{E}_{\text{CVB},2} = \frac{1}{\sqrt{2}} \left[\vec{e}_H \cdot \text{LG}^h_{0,0}\, \text{e}^{\text{i}\varphi} + \vec{e}_V \cdot \text{LG}^h_{0,0}\, \text{e}^{-\text{i}\varphi} \right].$$

The QWP_2 subsequently converts linear (horizontal and vertical) into circular (right- and left-handed) polarization basis, i.e.

$$\vec{E}_{\text{CVB},3} = \frac{1}{\sqrt{2}} \left[\vec{e}_R \cdot \text{LG}^h_{0,0}\, \text{e}^{\text{i}\varphi} + \vec{e}_L \cdot \text{LG}^h_{0,0}\, \text{e}^{-\text{i}\varphi} \right].$$

Finally, after filtering the zeroth order in the $4f$-system, the amplitude naturally forms the typical donut shape by free space propagation, so that $\text{LG}^h_{0,0}\, \text{e}^{\pm\text{i}\varphi}$ is approximately translated into $\text{LG}^h_{0,\pm 1}$.[24] Consequently, a light field according to Eq. (2.46) is sculpted.

Obviously, this approach is not limited to radial CVBs but can be used to realize any CVB following Eq. (2.26) by combining the multiplexing system with natural amplitude shaping.

2.4.4.5 Conclusion

Within this section a novel holographic single-beam method was proposed for the customization of fully-structured light. In contrast to known techniques for amplitude, phase and polarization modulation this cost-effective multiplexing approach facilitates the application of a standard non-HD SLM in full-screen mode allowing for optimal spatial resolution [86]. The functionality of the system was demonstrated experimentally by double-phase modulation as well as the customization of phase and polarization structured singular light fields. Furthermore, the inclusion of natural and on-demand amplitude structuring was proposed. Beyond, a modification of the multiplexing approach was outlined revealing the ability to tailor arbitrary CVBs by the developed experimental system.

References

1. Hawking SW, Penrose R (1970) The singularities of gravitational collapse and cosmology. Proc R Soc Lond A 314:529
2. Gottstein G (2013) Physical foundations of materials science. Springer Science & Business Media
3. Rubinsztein-Dunlop H et al (2017) Roadmap on structured light. J Opt 19:013001
4. Dennis MR, O'Holleran K, Padgett MJ (2009) Singular optics: optical vortices and polarization singularities. In: Wolf E (ed) Progress in optics, vol 53, Chap. 5. Elsevier, pp 293–363

[24] Precisely, $\text{LG}^h_{0,0}\, \text{e}^{\pm\text{i}\varphi}$ represent hypergeometric-Gaussian modes [102] also upon propagation, which have a broader transverse expansion than LG modes. However, approximately LG fields are formed.

5. Soskin MS, Vasnetsov MV (2001) Singular optics. In: Wolf E (ed) Progress in optics, vol 42, Chap. 4. Elsevier, pp 219–277
6. Dennis MR (2001) Topological singularities in wave fields. PhD thesis. University of Bristol
7. Freund I (2001) Polarization flowers. Opt Commun 199:47
8. Nye JF (1997) Line singularities in wave fields. English. Philos Trans: Math Phys Eng Sci 355: 2065
9. Hunt BR, Kaloshin VY (2010) Prevalence. In: Broer H, Hasselblatt B, Takens F (eds) Handbook of dynamical systems, vol 3, Chap. 2. Elsevier Science, pp 43–87
10. Oxtoby JC (2013) Measure and category: a survey of the analogies between topological and measure spaces, vol 2. Springer Science & Business Media
11. Dennis MR (2002) Polarization singularities in paraxial vector fields: morphology and statistics. Opt Commun 213:201
12. Saleh BEA, Teich MC (1991) Fundamentals of photonics. Wiley, New York
13. Götte JB, Barnett SM (2012) Light beams carrying orbital angular momentum. In: Andrews DL, Babiker M (eds) The angular momentumm of light, Chap. 1. Cambridge University Press, pp 1–30
14. Goodman J (2008) Introduction to Fourier optics. McGraw-Hill
15. Roux F (2014) Optical waves. In: Forbes A (ed) Laser beam propagation - generation and propagation of customized light, Chap. 1. CRC Press, p. 3–40
16. Kogelnik H, Li T (1966) Laser beams and resonators. Appl Opt 5:1550
17. Durnin J, Miceli JJ, Eberly JH (1987) Diffraction-free beams. Phys Rev Lett 58:1499
18. Gori F, Guattari G, Padovani C (1987) Bessel-Gauss beams. Opt Commun 64:491
19. Berry M, Dennis M (2001) Polarization singularities in isotropic random vector waves. Proc R Soc Lond A 457:141
20. Nye JF (1999) Natural focusing and fine structure of light: caustics and wave dislocations. CRC Press
21. Nye JF, Berry MV (1974) Dislocations in wave trains. Proc R Soc Lond A 336:165
22. Berry MV (1981) Singularities in waves and rays. In: Kléman M (ed) Balian J-PPR. North-Holland, Les Houches Session XXV - Physics of Defects
23. Allen L et al (1992) Orbital angular-momentum of light and the transformation of Laguerre-Gaussian laser modes. Phys Rev A 45:8185
24. Padgett MJ, Allen L (1995) The poynting vector in Laguerre-Gaussian laser modes. Opt Commun 121:36
25. Andrews DL, Babiker M (2012) The angular momentum of light. Cambridge University Press
26. Wisniewski-Barker E, Padgett MJ (2015) Orbital angular momentum. In: Andrews DL (ed) Photonics - fundamentals of photonics and physics, vol 1, Chap. 10. Wiley, pp 321–340
27. Berry MV, Dennis MR (2001) Knotted and linked phase singularities in monochromatic waves. Proc R Soc Lond A 457:2251
28. Otte E, Alpmann C, Denz C (2018) Polarization singularity explosions in tailored light fields. Laser Photon Rev 12:1700200
29. Siegman A (1986) Lasers. University Science Books
30. Boyd GD, Gordon JP (1969) Confocal multimode resonator for millimeter through optical wavelength masers. Bell Syst Tech J 40:489
31. Schutza A (2010) Transmission of quantum information via Laguerre-Gaussian modes. McNair Schol J 14:8
32. Strogatz SH (2001) Nonlinear dynamics and chaos: with applications to physics, biology and chemistry. Perseus publishing
33. Gibson G et al (2004) Free-space information transfer using light beams carrying orbital angular momentum. Opt Express 12:5448
34. Mirhosseini M et al (2015) High-dimensional quantum cryptography with twisted light. New J Phys 17:033033
35. Mair A et al (2001) Entanglement of the orbital angular momentum states of photons. Nature 412:313

36. He H et al (1995) Direct observation of transfer of angular momentum to absorptive particles from a laser beam with a phase singularity. Phys Rev Lett 75:826
37. Friese M et al (1996) Optical angular-momentum transfer to trapped absorbing particles. Phys Rev A 54:1593
38. Plick WN et al (2013) Quantum orbital angular momentum of elliptically symmetric light. Phys Rev A 87:033806
39. Allen L, Barnett SM, Padgett MJ (2003) Optical angular momentum. CRC Press
40. Van Enk S, Nienhuis G (2001) Eigenfunction description of laser beams and orbital angular momentum of light. Opt Angular Moment 36
41. Padgett M, Bowman R (2011) Tweezers with a twist. Nat Photon 5:343
42. Berry MV, Dennis MR, Jr RLL (2004) Polarization singularities in the clear sky. New J Phys 6:162
43. Zhan Q (2009) Cylindrical vector beams: from mathematical concepts to applications. Adv Opt Photon 1:1
44. Biss DP, Youngworth KS, Brown TG (2006) Dark-field imaging with cylindricalvector beams. Appl Opt 45:470
45. Ndagano B et al (2017) Characterizing quantum channels with non-separable states of classical light. Nat Phys 13:397
46. Nivas JJ et al (2017) Surface structuring with polarization-singular femtosecond laser beams generated by a q-plate. Sci Rep 7:42142
47. Nye J (1983) Monstars on glaciers. J Glaciol 29:70
48. Huard S (1997) Polarization of light. Wiley
49. Wolf E (1959) Coherence properties of partially polarized electromagnetic radiation. Il Nuovo Cimento 13:1165
50. Born M, Wolf E (1999) Principles of optics: electromagnetic theory of propagation, interference and diffraction of light. CUP Archive
51. Galvez EJ (2015) Light beams with spatially variable polarization. In: Photonics - fundamentals of photonics and physics, vol 1, Chap. 3. Wiley, pp 61–76
52. Beckley AM, Brown TG, Alonso M (2010) Full Poincaré beams. Opt Express 18:10777
53. Milione G et al (2011) Higher-order Poincaré sphere, stokes parameters, and the angular momentum of light. Phys Rev Lett 107:053601
54. Marrucci L, Manzo C, Paparo D (2006) Optical spin-to-orbital angular momentum conversion in inhomogeneous anisotropic media. Phys Rev Lett 96:163905
55. Galvez EJ, Khadka S (2012) Poincaré9 modes of light. Proc SPIE 8274:82740Y–82740Y
56. Alpmann C et al (2017) Dynamic modulation of Poincaré9 beams. Sci Rep 7:8076
57. Maurer C et al (2007) Tailoring of arbitrary optical vector beams. New J Phys 9:78
58. Freund I (2002) Polarization singularity indices in Gaussian laser beams. Opt Commun 201:251
59. Freund I et al (2002) Stokes singularity relations. Opt Lett 27:545
60. Hall DG (1996) Vector-beam solutions of Maxwell's wave equation. Opt Lett 21:9
61. Zhan Q (2014) Vector beams. In: Forbes A (ed) Laser beam propagation: generation and propagation of customized light, Chap. 9. CRC Press, pp. 239–271
62. Otte E, Alpmann C, Denz C (2016) Higher-order polarization singularitites in tailored vector beams. J Opt 18:074012
63. Freund I (2001) Vortex flowers. Opt Commun 196:63
64. Freund I (2001) Poincaré vortices. Opt Lett 26:1996
65. Mermin ND (1979) The topological theory of defects in ordered media. Rev Mod Phys 51:591
66. Dennis MR (2011) Fermionic out-of-plane structure of polarization singularities. Opt Lett 36:3765
67. Galvez EJ et al (2014) Generation of isolated asymmetric umbilics in light's polarization. English Phys Rev A 89:3
68. Otte E, Denz C (2018) Sculpting complex polarization singularity networks. Opt Lett 43:5821
69. Garcia-Gracia H, Gutiérrez-Vega JC (2016) Polarization singularities in nondiffracting Mathieu-Poincaré beams. J Opt 18:014006

70. Dudley A et al (2013) Generating and measuring nondiffracting vector Bessel beams. Opt Lett 38:3429
71. Otte E et al (2018) Recovery of nonseparability in self-healing vector Bessel beams. Phys Rev A 98:053818
72. Gibson CJ et al (2018) Control of polarization rotation in nonlinear propagation of fully structured light. Phys Rev A 97:033832
73. Davis JA et al (1999) Encoding amplitude information onto phase-only filters. Appl Opt 38:5004
74. Andrews DL (2011) Structured light and its applications: An introduction to phasestructured beams and nanoscale optical forces. Academic
75. Woerdemann M et al (2013) Advanced optical trapping by complex beam shaping. Laser Photon Rev 7:839
76. Han W et al (2013) Vectorial optical field generator for the creation of arbitrarily complex fields. Opt Express 21:20692
77. Hao J et al (2014) Light field shaping by tailoring both phase and polarization. Appl Opt 53:785
78. Rong Z-Y et al (2014) Generation of arbitrary vector beams with cascaded liquid crystal spatial light modulators. Opt Express 22:1636
79. Chen Z et al (2015) Complete shaping of optical vector beams. Opt Express 23:17701
80. Otte E et al (2015) Complex light fields enter a new dimension: holographic modulation of polarization in addition to amplitude and phase. Proc SPIE 9379:937908–937908
81. Rosales-Guzmán C, Bhebhe N, Forbes A (2017) Simultaneous generation of multiple vector beams on a single SLM. Opt Express 25:25697
82. Xie Y-Y et al (2015) Simple method for generation of vector beams using a small-angle birefringent beam splitter. Opt Lett 40:5109
83. Galvez EJ et al (2012) Poincaré-beam patterns produced by nonseparable superpositions of Laguerre-Gauss and polarization modes of light. Appl Opt 51:2925
84. Liu S et al (2018) Highly efficient generation of arbitrary vector beams with tunable polarization, phase, and amplitude. Photon Res 6:228
85. Otte E et al (2018) Entanglement beating in free space through spin-orbit coupling. Light Sci Appl 7:18009
86. Otte E, Tekce K, Denz C (2018) Spatial multiplexing for tailored fully-structured light. J Opt 20:105606
87. Otte E, Asché E, Denz C (2019) Shaping optical spin flow topologies by the translation of tailored orbital phase flow. J Opt 21:064001
88. Liesener J et al (2000) Multi-functional optical tweezers using computer-generated holograms. Opt Commun 185:77
89. Rosales-Guzmán C et al (2017) Multiplexing 200 spatial modes with a single hologram. J Opt 19:113501
90. Wulff KD et al (2006) Aberration correction in holographic optical tweezers. Opt Express 14:4169
91. Hell SW, Wichmann J (1994) Breaking the diffraction resolution limit by stimulated emission: stimulated-emission-depletion fluorescence microscopy. Opt Lett 19:780
92. Schaefer B et al (2007) Measuring the Stokes polarization parameters. Am J Phys 75:163
93. Boivin A, Dow J, Wolf E (1967) Energy flow in the neighborhood of the focus of a coherent beam. JOSA 57:1171
94. Vasnetsov M et al (2000) Wavefront motion in the vicinity of a phase dislocation:"optical vortex". Opt Spectrosc 88:260
95. Schouten HF et al (2004) The diffraction of light by narrow slits in plates of different materials. J Opt A: Pure Appl Opt 6:S277
96. Bekshaev AY, Soskin M (2007) Transverse energy flows in vectorial fields of paraxial beams with singularities. Opt Commun 271:332
97. Bekshaev A, Bliokh KY, Soskin M (2011) Internal flows and energy circulation in light beams. J Opt 13:053001

98. Tabor M (1989) Chaos and integrability in nonlinear dynamics: an introduction. Wiley
99. Gbur GJ (2017) Singularities of the poynting vector. In: Pike ER, Brown RGW (eds) Singular optics, vol 1, Chap. 8. CRC Press, pp 259–292
100. Cardano F et al (2012) Polarization pattern of vector vortex beams generated by q-plates with different topological charges. Appl Opt 51:C1
101. Esseling M (2014) Photorefractive optoelectronic tweezers and their applications. Springer
102. Karimi E et al (2007) Hypergeometric-Gaussian modes. Opt Lett 32:3053

Chapter 3
Vectorial Light Fields and Singularities in 3d Space

In the previous chapter, singular light fields were considered in 2d space, revealing striking features as the relation between singularity indices and spatial shape of the respective light field or the ability to customize not only amplitude and phase but also polarization and even energy flow with respective singularities. Beyond their transverse appearance and characteristics, in particular the longitudinal evolution in 3d space, hence the propagation dynamics of structured singular light fields reveals unique findings, especially with respect to scalar and vectorial singularity dynamics (e.g., see Refs. [1–5]). For instance, scalar phase singularities may form 3d vortex links and knots paving the way to future applications in laser technology or optical trapping [1, 2, 6]. Moreover, polarization singularities enable even richer topologies due to their vectorial character [7]. Already simple configurations formed by q-plates [8], the off-axis superposition of orthogonally polarized optical vortices [9], or superimposed cylindrical vector beams (CVBs) [10] have proven the diversity of complex polarization and singularity topologies in 3d space. Beyond, by the application of, e.g., the Pockels effect [11], apertures [12], or anisotropic, birefringent media [13] the manipulation of singularity evolution was shown.

Although already simple field configurations enable interesting findings in the field of singular optics, more complex, even tailored configurations embedding, e.g., multiple singularities have not been considered so far. Furthermore, the 3d behavior of singular light is also of particular interest considering applications. In these, not only specific transverse 2d planes of light can be chosen, but the complete light field and singularities in 3d space play a crucial role. This fact highlights the demand for structured singular fields adaptable in 3d space to specific applications. However, till now, the on-demand control of 3d propagation dynamics of structured light and especially its singularities could not be realized.

In this chapter vectorial fields imparting scalar and/or vectorial singularities are tailored and studied in 3d space exhibiting not yet known controllable propagation

E. Otte, *Structured Singular Light Fields*, Springer Theses,
https://doi.org/10.1007/978-3-030-63715-6_3

dynamics and even complete new beam classes. Besides demonstrating the customization of complex 2d singularity networks being conserved upon propagation in 3d space (Sect. 3.1), the on-demand control of singularity propagation dynamics is evinced. On the one hand non-generic scalar and vectorial singularities are combined, resulting in an interaction of propagation dynamics forming polarization singularity "explosions" (Sect. 3.2). On the other hand, non-diffracting and even self-imaging vectorial fields are designed embedding complex singularity configurations inheriting non-diffracting and self-imaging properties (Sect. 3.3).

Note that the results presented in this chapter have in parts been published in Refs. [14–16].

3.1 Sculpting Structurally Stable Vector Fields and Conserved Singularity Networks

Naturally, optical singularities appear in arrangements of multiple singular points as in the blue daylight sky or in optical speckle fields [17, 18]. Hence, the in-depth evaluation of these configurations, especially in 3d space, is fundamentally meaningful since it contributes to the field of singular optics as well as fundamental understanding of wave phenomena. However, current fundamental as well as applied research is mainly focused on basic light fields as CVBs including single on-axis singularities. In contrast, investigations on networks of multiple singularities—"singularity networks" (SNs)—is still in its infancy, with only a few studies on SNs created by, e.g., multi-wave interference [19–22], birefringent media [23, 24], or coherence control [25]. Obviously, in these cases the customization of SNs is limited. Crucially, respective light fields do not propagate in a structurally stable way for arbitrary distances in 3d space, which hampers its intended application, e.g., in free space communication. Here, a novel class of structurally stable vectorial light fields is proposed [16] which does not only include a single singularity, but a complex customizable SN, which is conserved upon propagation.

3.1.1 *Idealized Propagation of CVBs*

As demonstrated in Sects. 2.3.2 and 2.4.2, vectorial light fields can be generally seen as a superposition of orthogonally polarized scalar light fields. Following this principle, well-known CVBs (cf. Sect. 2.3.2) are composed of LG modes of orthogonal spatial shape. The respective cylindrical symmetry of LG modes induces the occurrence of an individual on-axis vectorial point singularity in CVBs. Further, as CVBs consist of LG modes, they feature structurally stable propagation behavior, as visualized in Fig. 3.1. Here, Fig. 3.1a illustrates the direction of propagation of the light field ($+z$-direction) and the plane which is named 0-plane within

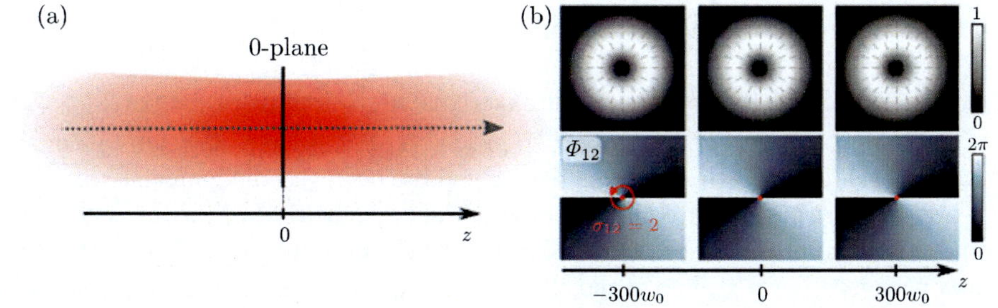

Fig. 3.1 Structurally stable radial CVB upon propagation (idealized case). **a** Illustration of 0-plane at $z = 0$ of considered light field. **b** Polarization structure upon intensity and Stokes field phase Φ_{12} of a propagating radially polarized CVB; $z = \{-300w_0, 0, 300w_0\}$ with small divergence angle $\theta_d = 0.0005°$

this thesis ($z = 0$). In Fig. 3.1b the calculated propagation (Eqs. (2.14), (2.26)) of a radially polarized CVB is visualized as a representative example. Obviously, the polarization configuration and intensity (top) does not change, as shown for a range of $z = [-300w_0, 300w_0]$ (w_0: fundamental Gaussian beam waist). Since a small divergence angle of $\theta_d = 0.0005°$ is assumed, no significant scaling in the beam's transverse size is observed. The propagation is studied with no perturbations considered, i.e., in an idealized case, so that, here, the single embedded (non-generic) V-point of index $\sigma_{12} = 2$ also propagates stably, as highlighted in Φ_{12} (marked red).

Besides being well-studied, due to their structural stability CVBs are of special interest for applications in information technologies in which singularities serve as integer information units [26, 27]. However, vectorial light fields could not reach its full potential yet, as configurations of multiple singularities in structurally stable fields have not been considered so far. Besides being valuable for fundamental research, the realization of SNs within the transverse plane of a light field could considerably increase the number of spatial as well as polarization based degrees of freedom (DoFs) for information processing.

3.1.2 Symmetry Breaking for Realizing Singularity Networks

Here, a more general class of structurally stable vector fields, namely "Ince-Gaussian vector modes" (IVMs) [16] is proposed. In contrast to CVBs these fields do not only include a single on-axis singularity but can embed an extended network of multiple singular points, thus, fulfill the above mentioned current demand for SNs. The network formation is enabled by breaking typically observed cylindrical symmetry via the application of elliptical symmetric Ince-Gaussian (IG) modes as basis within a superposition approach (cf. Eq. (2.26)).

3.1.2.1 Ince Gaussian Vector Modes

Scalar IG modes ($IG^{e,o,h}_{p,m,\varepsilon}$) represent exact analytic solutions of the paraxial wave equations in elliptical coordinates $\vec{r} = [\xi, \eta, z]^T$ [28]. Similar to LG beams, IG modes can be even (e), odd (o) or helical (h), whereby, here, the fundamental states [28] (e, o) are considered. In contrast to cylindrical symmetric LG modes with radial and azimuthal mode indices n and ℓ, respectively, IG beams depend on three mode numbers: its order p, degree m and ellipticity ε. Equations for fundamental IG modes can be summarized as

$$IG^j_{p,m,\varepsilon}(\vec{r}) = N^j \, \frac{w_0}{w(z)} \, P^j_{p,m}(\mathrm{i}\xi, \varepsilon) \, P^j_{p,m}(\eta, \varepsilon) \cdot \mathrm{e}^{\mathrm{i}p\phi^G(z)} \, E_G(\vec{r}), \qquad j = \{e, o\} \tag{3.1}$$

with N^j as normalization constant and $P^j_{p,m}$ representing the even and odd Ince polynomial [28]. Further, w_0 and $w(z)$ represent the beam waist and width, $E_G(\vec{r})$ is the fundamental Gaussian mode with its Gouy phase $\phi^G(z)$ (cf. Eq. (2.11)). Consider that the ellipticity ε enables the inclusion of LG as well as Hermite-Gaussian (HG) modes in the IG class, since for $\varepsilon = 0$ ($\varepsilon = \infty$) LG (HG) modes are derived. Hence, the IG class embodies an overall class for these Gaussian modes.

By incoherently combining different IG modes the class of IVMs is created [16], following the equation

$$\vec{IVM} = \frac{1}{\sqrt{2}} \left[\vec{IG}^{x,j_x}_{p_x,m_x,\varepsilon_x} + \vec{IG}^{y,j_y}_{p_y,m_y,\varepsilon_y} \right]. \tag{3.2}$$

Here, x, y indicate the affiliation to horizontal ($\vec{e}_x = [1, 0]^T$) and vertical ($\vec{e}_y = [0, 1]^T$) linear polarization, respectively. Thus, $\vec{IG}^{x,y} = \vec{e}_{x,y} \cdot IG^{x,y} \cdot \mathrm{e}^{\mathrm{i}\alpha_{x,y}}$ ($\alpha_x = -\alpha_y$). Note that besides linear polarization basis $\vec{e}_{x,y}$ and fundamental (e, o) IG modes, different basis states as circular or diagonal polarizations and helical IG modes could be applied as well. Hence, depending on chosen basis states and mode indices, also CVBs following Eq. (2.26) can be formed.

An example of IVMs is presented in Fig. 3.2 with numerical (top) and experimental (bottom) results in Fig. 3.2a for $z = 0$. Here, a horizontally polarized $IG^{x,e}_{4,4,0}$ and vertically polarized $IG^{y,o}_{4,2,5}$ (mode numbers in Fig. 3.2b; $\alpha_{x,y} = 0$ here and in the following) are combined, whose intensity $I_{x,y}$ and phase $\phi_{x,y}$ are shown in Fig. 3.2c. While numerical results are calculated from Eqs. (3.1) and (3.2), the respective IVM is formed experimentally by joining amplitude and phase modulation (Sect. 2.4.1) and a DPMS ($\beta_{1,2} = -\pi/4$, Sect. 2.4.3) (details see Ref. [29] or Sect. 3.3.2). First, a SLM modulates the amplitude and global phase[1] of the IVM, second, this scalar field enters the DPMS and the respective polarization structure is imprinted. The resulting vector field is measured by a Stokes parameter measurement system (Sect. 2.4.2). Note that, even though an interferometric approach (Sect. 2.4.2) might be the more

[1]The global phase represents the phase being imprinted onto the whole light field independent of the polarization structure. Hence, for a light field $\vec{E} = \vec{e} \cdot E_0 \cdot \exp(\mathrm{i}\phi)$, $\vec{e}$ is the normalized Jones vector (polarization), E_0 the global amplitude, and ϕ the global phase.

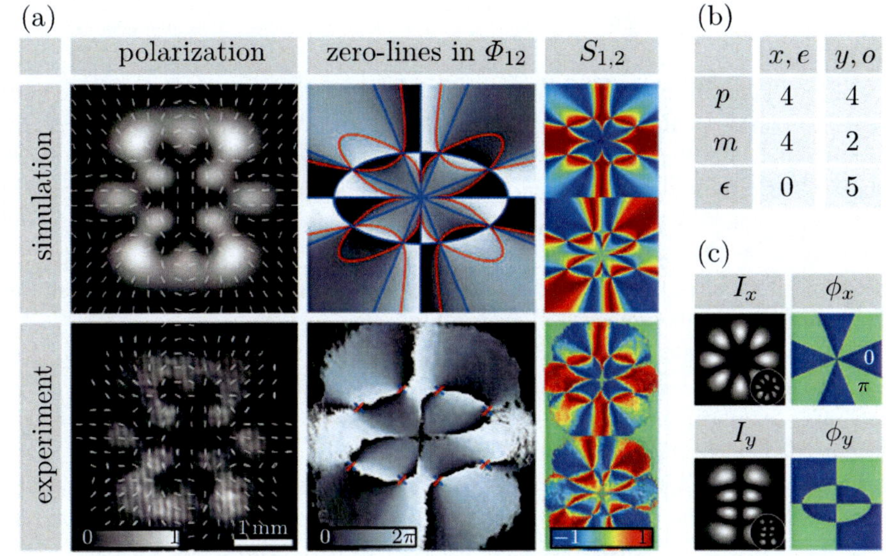

Fig. 3.2 Illustrative example of IVMs in simulation and experiment (top and bottom in **a**, $z = 0$). **a** Polarization structure on intensity/Stokes parameter s_0, (intersections of) zero-lines ($Z_{\Re,\Im}$: red, blue) in phase Φ_{12} and Stokes parameters $S_{1,2}$ (top, bottom) of exemplary IVM. **b** Parameters as well as **c** intensity $I_{x,y}$ (inset: experiment) and phase $\phi_{x,y}$ of combined scalar $\mathrm{IG}^{x,y}$ modes for the formation of the IVM. [Adapted with permission from Ref. [16]; © 2018 The Optical Society]

intuitive system, this APP modulation setup has the distinct advantage of avoiding errors within the IVM originating from non-perfect superposition of orthogonally polarized IG modes.

Clearly visible in Fig. 3.2a, experimental results perfectly match respective simulations. Here, for both, simulation and experiment, the linear polarization structure upon the total intensity (= Stokes parameter s_0), the phase Φ_{12} with respective zero-lines $Z_{\Im,\Re}$ (blue, red), and Stokes parameters $S_{1,2}$ (top, bottom; $S_3 = 0$) are presented. A complex linear polarization configuration is created which contains an assembly of eight V-point singularities ($\sigma_{12} = \pm 2$), identified as intersections of zero-lines. Consider that, in the experiment, these non-generic singularities might be splitted into two very close C-points due to experimental constraints. This network of singularities is enabled by the non-cylindrical, but elliptical symmetry of IG^y, hence, by breaking the typical cylindrical symmetry. Accordingly, singularities are located off-axis on an elliptical ring. Thereby adjacent V-points are of opposite index sign or an extremum/saddle point (self-intersection of zero-lines) is found between them, so that the sign rule (cf. Sect. 2.1) is kept. Note that on-axis, red $Z_\Re$ zero-lines do not intersect with blue $Z_\Im$ lines. Hence, the self-intersection of $Z_\Im$ lines mark an extremum.

3.1.2.2 Variety of IVMs

Through the ability to adapt all parameters of the IVM, especially the mode numbers p, m and ε of $\mathrm{IG}^{x,y}$ independently, a huge variety of IVMs and, thus, of SNs becomes available [16]. In Fig. 3.3 an insight into this variety is given by some exemplary IVMs (simulations, $z = 0$, no perturbations) highlighting the customizability of SNs. Per row (I-VII), Fig. 3.3a includes the intensity and phase of IG modes (top, bottom: $\mathrm{IG}^{x,y}$), the polarization configuration on s_0, the phase Φ_{12} with respective zero-lines ($Z_{\mathfrak{R},\mathfrak{I}}$: red, blue), and normalized Stokes parameters $S_{1,2}$ (top, bottom). The respective mode indices are listed in Fig. 3.3b.

Above, it was stated that IVMs also include CVBs according to Eq. (2.26), which is now proven in row I by the example of an IVM constituted of $\mathrm{IG}^{x,e}_{0,2,4}$ and $\mathrm{IG}^{y,o}_{0,2,4}$, both including an ellipticity of zero. By the incoherent superposition of these cylindrical symmetric beams a single on-axis singularity, namely, a V-point of index $\sigma_{12} = 4$ is formed. If the ellipticity is increased, the intended symmetry breaking is initiated, as visualized by the example in row II. Here, mode numbers $m = m_x = m_y$ and $p = p_x = p_y$ are not changed in comparison to row I, however, the increase of ellipticity to $\varepsilon = \varepsilon_x = \varepsilon_y = 5$ causes the creation of a complex polarization structure of elliptical symmetry (cf. $S_{1,2}$) embedding a SN. The SN is constituted of ten V-points ($\sigma_{12} = \pm 2$) with six woven saddle points, keeping the sign rule. Singularities and saddles are located on elliptical rings or hyperbolic lines formed by $Z_{\mathfrak{I},\mathfrak{R}}$, which mirrors the symmetry of applied $\mathrm{IG}^{x,y}$ basis modes.

Row III presents an example in which $m_x \neq m_y$. By this adaption, the number of polarization singularities is increased to 12, forming almost a square lattice due to the hyperbolic symmetry of ϕ_x. Still the order of singularities is kept at $\sigma_{12} = \pm 2$ each. Besides choosing $m_{x,y}$ independently, one can also set $\varepsilon_{x,y}$ to different values. In row IV, mode numbers $p_x = p_y$ and $m_x = m_y$, while $\varepsilon_x = 0$, i.e., it is a LG mode, and $\varepsilon_y = 5$. These settings result in an arrangement of four lowest-order off-axis V-points ($\sigma_{12} = \pm 2$) and a higher-order on-axis singularity with $\sigma_{12} = \pm 4$. The latter one originates from the cylindrical symmetry of IG^x, representing a LG mode, whereas the elliptical symmetry of IG^y causes the occurrence of additional off-axis singularities. Note that intersections of blue elliptical and circular ring of $Z_\mathfrak{I}$ do not mark V-points but saddles, since red $Z_\mathfrak{R}$ do not cross here.

Of course, one can also choose to set both, $m_{x,y}$ and $\varepsilon_{x,y}$, independently. This is shown by the example in row V, reprinting the results in Fig. 3.2. Here, a SN of eight singular points is found, completed by an on- and four off-axis saddle points for the formation of a continuous polarization structure. Beyond, all mode indices can be defined independently, as exemplified in row VI. In this case, a SN of eight V-points is sculpted. Additionally, eleven saddles are embedded, one of them on-axis. Note that crossing points of blue circular and elliptical $Z_\mathfrak{I}$ ring again mark saddles not V-points (cf. $S_{1,2}$).

As mode numbers can be chosen as desired, various singularity configurations are enabled, not only located on a single ring but even on multiple ones, facilitating SNs embedding a large number of singularities. This is visualized by another example in row VII. By increasing the ellipticity of $\mathrm{IG}^{x,y}$ with $\varepsilon_{x,y} \neq 0$, a SN of 28 V-points with

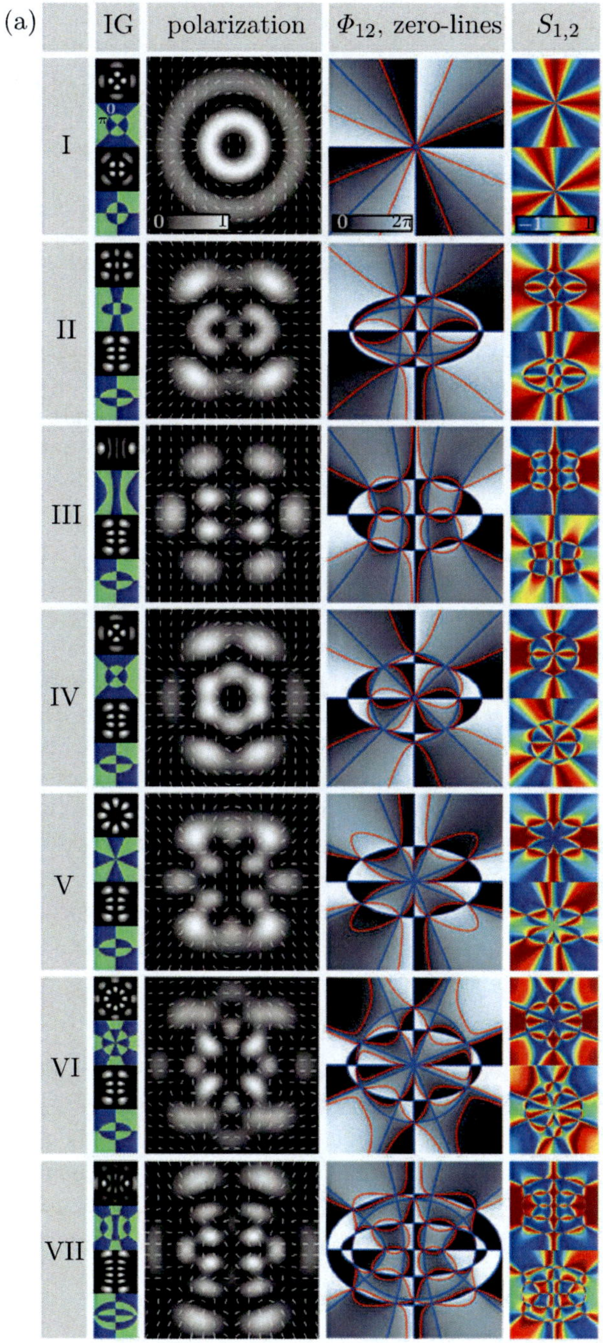

(b)

		x, e	y, o
I	ϵ	0	0
	m	2	2
	p	4	4
II	ϵ	5	5
	m	2	2
	p	4	4
III	ϵ	5	5
	m	4	2
	p	4	4
IV	ϵ	0	5
	m	2	2
	p	4	4
V	ϵ	0	5
	m	4	2
	p	4	4
VI	ϵ	0	5
	m	4	2
	p	6	4
VII	ϵ	2	7
	m	4	2
	p	6	6

Fig. 3.3 Insight into the variety of IVMs with different mode numbers per row I-VII ($z = 0$). **a** Intensity and phase of superimposed IG modes (top, bottom: $\mathrm{IG}^{x,y}$), polarization structure on total intensity (s_0), zero-lines ($Z_{\Re,\Im}$: red, blue) in phase Φ_{12}, and Stokes parameters $S_{1,2}$ (top, bottom; $S_3 = 0$) of IVMs. **b** Mode indices per row. [Adapted with permission from Ref. [16]; © 2018 The Optical Society]

woven saddle points is tailored within a complex extended polarization structure (cf. $S_{1,2}$, Φ_{12}).

Hence, by the adaption of mode indices p, m and ε manifold IVMs embedding sophisticated SNs can be shaped. Thereby, the number, arrangement and order of included singularities can be tailored on-demand by the choice of indices.

3.1.3 Structural Stability and Propagation Dynamics

As superimposed modes $\mathrm{IG}_{x,y}$ are both exact analytic solutions of the paraxial wave equation, thus, propagate structurally stably, one could assume that their superposition, i.e., an IVM, inherits this property. However, as IG modes represent higher-order Gaussian modes, the respective Gouy phase of each, IG^x and IG^y, needs to be considered as this phase depends on the Gaussian modes indices [28]. For IG beams, the mode index dependent Gouy phase is defined as

$$\phi^G_{p,m,\varepsilon}(z) = (1+p)\,\phi^G(z) \tag{3.3}$$

with the Gouy phase $\phi^G_{0,0,0}(z) = \phi^G(z)$ of the fundamental Gaussian beam. Consequently, if IG modes of different order $p_x \neq p_y$ are incoherently superimposed, a varying phase difference between IG_x and IG_y is acting upon propagation. Thus, the respective polarization structure of the IVM is changing in z-direction, i.e., it shows a structural unstable propagation behavior. Vice versa, for the realization of a structurally stable IVM, orders p_x and p_y need to be equal [16], i.e., $p_x = p_y$, whereas the degrees $m_{x,y}$ and ellipticities $\varepsilon_{x,y}$ can be set independently. By this approach, the phase difference between IG modes does not change upon propagation so that their individual structural stability results in the structural stability of the composite IVM.

3.1.3.1 Numerical Approach

In the following, the propagation dynamics of IVMs is studied by some examples. For this purpose, the IVM propagation in free space as well as under turbulence is investigated, whereby the behavior of embedded SNs is considered as well. For the calculation of respective propagation dynamics, a spectral numerical method [30] is applied.[2] Thereby, the propagated light field $\vec{E}(x,y,z) = [E_x(x,y,z), E_y(x,y,z)]^T = \vec{\mathrm{IVM}}(x,y,z)$ (with $\vec{\mathrm{IVM}}(\xi,\eta,z) \rightarrow \vec{\mathrm{IVM}}(x,y,z)$ by coordinate transformation) is represented by the components [30]

$$E_{x,y}(x,y,z) = \mathrm{e}^{\mathrm{i}z\nabla^2_{\perp}E_{x,y}(x,y,0)/2}\,E_{x,y}(x,y,0). \tag{3.4}$$

[2]Even though the idealized free space propagation could be calculated from Eqs. (3.1) and (3.2), the numerical approach is applied. This is due to keeping the results for free space and turbulence propagation comparable, as the latter one requires a numerical technique for its determination.

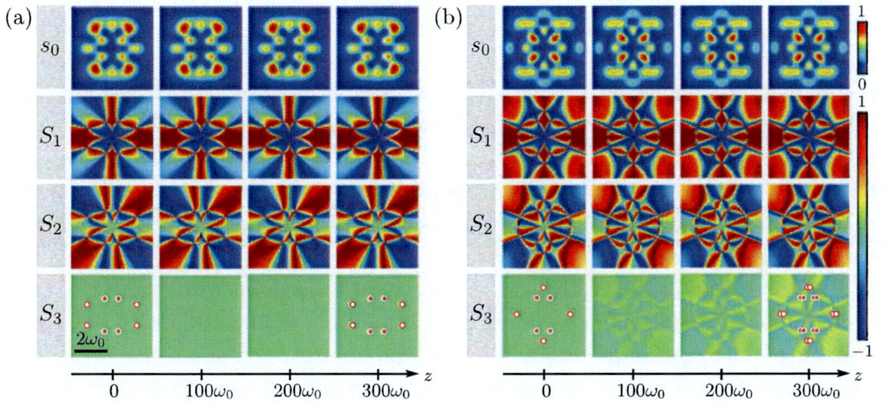

Fig. 3.4 Free space propagation of **a** structurally stable and **b** unstable IVMs with $\vec{\mathrm{IG}}^{x,e}_{4,4,0} + \vec{\mathrm{IG}}^{y,o}_{4,2,5}$ and $\vec{\mathrm{IG}}^{x,e}_{6,4,0} + \vec{\mathrm{IG}}^{y,o}_{4,2,5}$, respectively. Top to bottom: Stokes parameters s_0 and $S_{1,2,3}$ per propagation distance $z = \{0, 100, 200, 300\}w_0$ (w_0: fundamental Gaussian beam waist). Negatively/positively indexed singularities indicated by red/white circles, filled white/red. [Adapted with permission from Ref. [16]; © 2018 The Optical Society]

with $\nabla^2_\perp = \partial^2/\partial x^2 + \partial^2/\partial y^2$ and $E_{x,y}(x, y, 0)$ representing the light field distribution in the 0-plane ($z = 0$, cf. Fig. 3.1). If the (inverse) Fourier transform $\mathcal{F}^{(-1)}$ is utilized, the differential operator can be replaced according to $\nabla^2_\perp \rightarrow -(k_x^2 + k_y^2)$, with $k_{x,y}$ being the coordinates in Fourier space ($\vec{k} = [k_x, k_y, k_z]^T$: wave vector), so that the propagated light field can be numerically computed by [30]

$$E_{x,y}(x, y, z) = \mathcal{F}^{-1}\left[\mathrm{e}^{-\mathrm{i}(k_x^2+k_y^2)z/2}\,\mathcal{F}\left[E_{x,y}(x, y, 0)\right]\right]. \tag{3.5}$$

3.1.3.2 Free Space Propagation

For the visualization of structurally stable and unstable behavior, the propagation of an IVM with $p_x = p_y$ is compared to the one of an IVM with $p_x \neq p_y$. Precisely, the propagation dynamics of a structurally stable IVM constituted of $\vec{\mathrm{IG}}^{x,e}_{4,4,0} + \vec{\mathrm{IG}}^{y,o}_{4,2,5}$ (Fig. 3.3, row V) and an unstable one with $\vec{\mathrm{IG}}^{x,e}_{6,4,0} + \vec{\mathrm{IG}}^{y,o}_{4,2,5}$ (Fig. 3.3, row VI) are investigated, first, upon free space propagation (divergence angle $\approx 0.0005°$; no perturbations[3]). Results are shown in Fig. 3.4 presenting the Stokes parameters s_0 and $S_{1,2,3}$ for different propagation distances $z = \{0, 100, 200, 300\}w_0$ for (a) the structurally stable and (b) the unstable case. For the structurally stable IVM, no

[3]Note that the numerical approach naturally includes minor perturbations caused by matrix calculations within the technique. However, the effect of these perturbations is insignificant here, but will reveal its advantage in the next section (Sect. 3.2) where it is applied to mimic the effect of SLM discretization.

changes are visible within the Stokes parameters, since $p_x = p_y$, so that no relative phase difference is formed. As a consequence, also polarization singularities, i.e., non-generic V-points ($\sigma_{12}^V = \pm 2$), in the respective SN are kept (white/red circles filled red/white on S_3: positively/negatively indexed singularities).

In contrast, in the structurally unstable case, it is $p_x \neq p_y$, thus, the two modes IG_x and IG_y inpart a z-dependent Gouy phase difference of $\Delta\phi^G(z) = 2\phi^G(z)$ (cf. Eq. (3.3)). This increasing difference causes a modification within the polarization structure upon propagation, as particularly observable in $S_3(z)$. As revealed by bluish and yellowish colors in S_3 (green: $S_3 = 0$), the former purely linearly polarized vector field is transforming into an ellipse field containing states of polarization with $S_3 \neq 0$. Hence, initial non-generic V-points in the vector field, being unstable under these perturbations, convert into generic singularities within the ellipse field—namely, C-points of pure circular polarization (cf. circles on S_3, Fig. 3.4b). V-points unfold into C-points as both correspond to Φ_{12}-singularities, the former one in a vector ($S_3 = 0$), the latter in an ellipse field ($S_3 \in [-1, 1]$), both characterized by the conserved quantity σ_{12} [18]. During this natural unfolding process, each V-point of index $\sigma_{12}^V = \pm N$ splits into N C-points of lowest order $\sigma_{12}^C = \pm 1$ (here: $N = 2$). Thus, the total singularity index is conserved with $\sigma_{12}^{\text{tot}} = M \cdot \sigma_{12}^V = M \cdot N \cdot \sigma_{12}^C$ for M splitting V-points. However, even though V-points unfold into C-points, the overall arrangement of singularities in the SN does not change upon propagation—only each V-point is substituted by two C-points, i.e., a SN of C-point pairs is formed (cf. circles on S_3, Fig. 3.4b). The C-points in each pair are spatially slightly separated from each other with increasing distance upon propagation, as singularities of the same index act repulsively to a certain degree [18]. Nevertheless, the distances of C-point pairs within the SN is significantly larger than the distance between the C-points constituting each pair, so that the SN structure can be considered as conserved.

Hence, for $p_x = p_y$ structurally stable IVMs with conserved V-point SNs are formed, while for $p_x \neq p_y$ IVMs of structural instability but with conserved SN of V-points in the 0-plane and C-points for $z \neq 0$ are created.

3.1.3.3 The Effect of Turbulence

To affirm the findings on the (in)stability of IVMs and the respective SNs, the propagation dynamics of these under major perturbations, namely, atmospheric turbulence [31, 32], are studied. This study also provides insight into the applicability of these innovative vectorial modes, e.g., in free space data communication, which requires structurally stable light fields even considering turbulence. Results are shown in Fig. 3.5.

To mimic propagation of modes through turbulent atmosphere, a disturbing phase distribution (single phase screen) representing a turbulence of strength $SR = 0.3$ (SR: Strehl ratio[4]) following a turbulence model based on Kolmogorov's theory [27,

[4]The Strehl ratio represents the ratio of mean irradiance from a point source at the receiver's position (on-axis), if turbulence is present, to the case without turbulence.

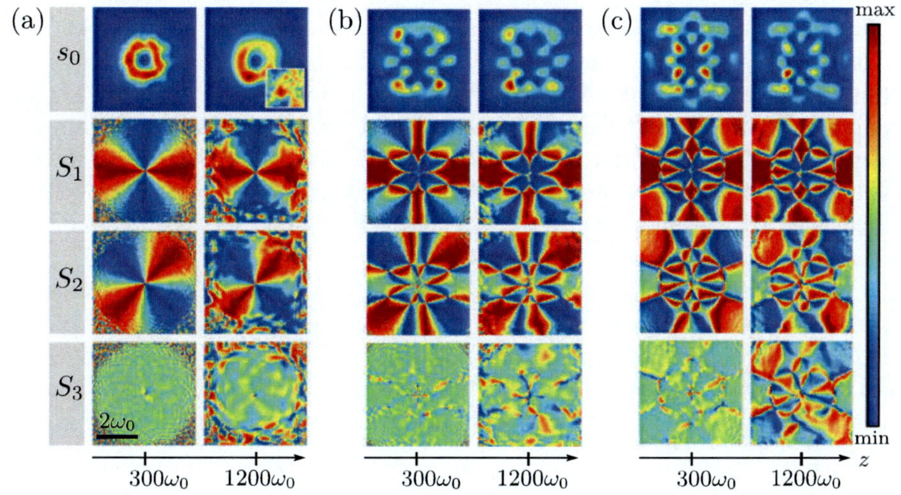

Fig. 3.5 Simulations on the effect of turbulence (inset in **a**, single phase screen at $z = 0$, $\in [0, \pi]$) on IVM propagation by the example of **a** a radial CVB, **b** structurally stable ($\vec{\mathrm{IG}}^{x,e}_{4,4,0} + \vec{\mathrm{IG}}^{y,o}_{4,2,5}$), and **c** unstable ($\vec{\mathrm{IG}}^{x,e}_{6,4,0} + \vec{\mathrm{IG}}^{y,o}_{4,2,5}$) IVM. Top to bottom: Stokes parameters $s_0 \in [0, 1]$ and $S_{1,2,3} \in [-1, 1]$ per propagation distance $z = \{300, 1200\}w_0$ (w_0: fundamental Gaussian beam waist). [Adapted with permission from Ref. [16]; © 2018 The Optical Society]

33, 34] is added to the initial beam in the 0-plane. The applied single phase screen is illustrated in Fig. 3.5a as an inset. First, as a representative example for CVBs, representing a subclass of IVMs, the propagation dynamics of a disturbed radial CVB is considered. CVBs are structurally stable and, due to this and their integer singularity index, have already served as integer information units [26]. Interestingly, it has been shown that CVBs are as resilient to turbulence as scalar modes of light [35], but due to their vectorial character broaden the number of accessible DoFs for information encoding. Here, the found resilience is approved by the study of Stokes parameters upon propagation considering turbulence, as visible in Fig. 3.5a. Parameters (top to bottom: s_0, $S_{1,2,3}$) are shown for propagation distances of $z = 300w_0$ and $1200w_0$, revealing no significant structural changes. The beam stays almost linearly polarized, only including states of minor ellipticity (cf. S_3). States of polarization imparting larger ellipticities are found close to the optical axis: Reacting to the perturbation originating from the turbulence, the non-generic V-point embedded on-axis in the CVB is unfolding upon propagation, converting into generic C-points enclosed by elliptical state of polarization. However, even though the V-point unfolds following singularity index conservation, the general structure of the CVB is kept, highlighting the feasibility of applying these modes for free space information encoding through turbulent atmosphere.

Additionally, Fig. 3.5b, c exemplify the propagation dynamics of the structurally stable and unstable IVMs of Fig. 3.4, respectively, if turbulence is considered [16]. For the (b) structurally stable IVM, similar effects as in the case of the CVB (a) are observed. Overall, the mode structure is conserved being almost linearly polarized. Non-linearly polarized states mainly appear at the former position of non-generic (thus, inherently unstable) V-points. Again, these non-generic perturbed singular points split into generic C-points of lowest order, causing the occurrence of elliptical states. In the (c) structurally unstable case, major changes within the polarization structure of the IVM are observed. Here, the mode clearly converts into an ellipse field, especially visible in S_3 for $z = 1200w_0$. Also in this case inherently unstable V-points unfold into generic C-points propagating stably. Nonetheless, in both cases, for the perturbed structurally stable and unstable IVM, the general arrangement of singularities is conserved, similar to split singularities in the structurally unstable IVM upon free space propagation. Hence, even though only structurally stable IVMs are suitable for free space communication considering turbulence, both, stable and unstable ones, reveal the not yet known conservation of SN upon propagation.

3.1.4 Conclusion

An overall class of vectorial light fields, namely Ince-Gaussian vector modes (IVMs) [16], was proposed which does not only include CVBs, but even all vectorial fields constituted of LG, HG or IG modes. By choosing IG modes as basis, the on-demand formation of complex singularity networks (SNs) is enabled by breaking the typically observed cylindrical symmetry of CVBs. This customization facilitates the study of fundamental properties of polarization singularities, which naturally occur in arrangements of multiple singular points as in the blue daylight sky. Beyond, these innovative modes pave the way to advanced free space data communication with a significantly increased number of DoFs, as IVMs can be tailored to be structurally stable or unstable, depending on chosen mode indices. The resilience of structurally stable IVMs was evinced in an idealized case as well as under major perturbations originating from turbulence. Additionally, the conservation of tailored SNs was demonstrated for both, the stable and unstable case—although the type of singularities may change from non-generic V-points in the 0-plane to generic, thus, stable C-points for $z \neq 0$, the overall arrangement of singularities is kept.

3.2 Polarization Singularity Explosions by Tailored Perturbation

It is well known that non-generic singularities unfold into their generic counterparts if they are perturbed. With respect to this conversion procedure, optical scalar singularities in the form of phase vortices are well-studied phenomena: under minor perturbations higher-order phase vortices naturally unfold into first-order ones upon propagation [17]. Thereby first-order singularities are located equally distant on a circle around the initial position of the higher-order singular point and twist around this position during propagation. Beyond, vectorial point singularities show a similar behavior, as evinced in the previous section. There, the stability of non-generic V-point singularities was studied, whereby singularities built networks and were embedded in modes of light constituting exact solutions of the wave equation. Proving the inherent instability of V-points, the splitting of V- into generic C-points was observed under perturbations of these modes.

Even though, separately, scalar as well as vectorial singularities have been investigated intensively (e.g., see Refs. [17, 36, 37] and refs. therein) with respect to unfolding properties, the combination of both in a joint light field is still in its infancy. Therefore, allowing for these missing essential insights into singular optics, in the following, non-generic V-points and higher-order phase singularities are combined in a slightly perturbed system to observe their interacting propagation dynamics [14]. Besides numerically investigating the combination of an individual V-point and a scalar singularity, a tailored joint light field is studied which embeds positively and negatively indexed polarization as well as phase singularities in very close vicinity, theoretically annihilating each other. Beyond, the genericity of line singularities is examined in a tailored ellipse field. By these studies the not yet observed control of the singularity splitting procedure as well as the transfer of splitting properties from one to another singularity class will be demonstrated.

3.2.1 *Polarization Singularity Splitting in Vector Fields*

3.2.1.1 Combining Scalar and Vectorial Singularities

In order to directly observe the properties of singularities itself, fields tailored purely in polarization and phase, i.e., not in amplitude, are used—thus, singularities are not positioned in extended dark areas as, e.g., in CVBs. As a consequence, these initial fields do not represent exact analytic solutions of the wave equation. The initial singular vector field with additional global phase variation at $z = 0$ is represented by

$$\vec{E} = \left[\cos\left(\frac{\sigma_{12}}{2}\cdot\varphi\right),\ \sin\left(\frac{\sigma_{12}}{2}\cdot\varphi\right)\right]^T \cdot \mathrm{e}^{\mathrm{i}\ell\varphi}\,. \tag{3.6}$$

The field[5] includes a V-point of index σ_{12} on axis superimposed with a phase singularity of topological charge ℓ. For studying the propagation dynamics of this field in 3d space, the spectral numerical approach presented in Sect. 3.1.3 is applied. This approach enables the inclusion of intended minor perturbations within the system by discretization: On the one hand, spatial discretization is imparted by the application of matrix calculations. On the other hand, phase values of the polar coordinate φ

[5] Note that this type of field could also be expressed as a superposition of circularly polarized hypergeometric-Gaussian modes [38] as outlined in Ref. [14].

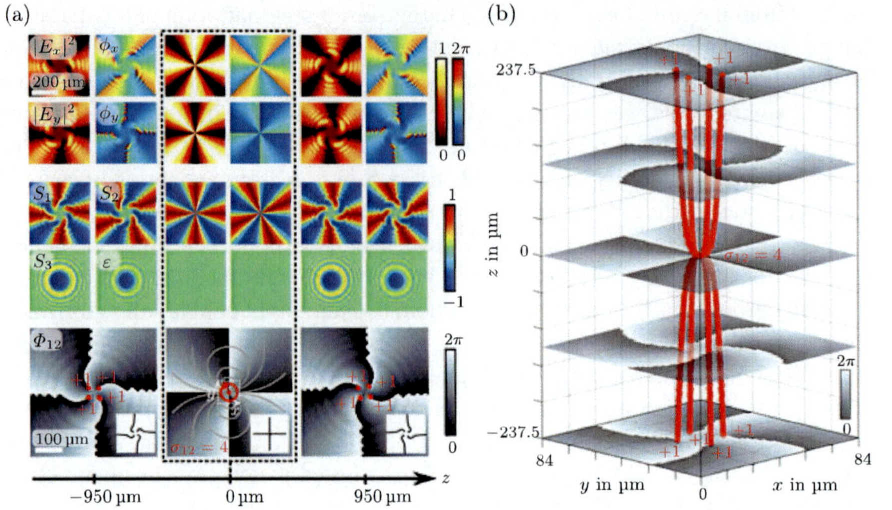

Fig. 3.6 Polarization singularity splitting by joining scalar and vectorial non-generic singularities. Shown are the numerically calculated propagation dynamics of a spider-shaped vector field with $\sigma_{12} = 4$ (cf. 0-plane) combined with a phase vortex of charge $\ell = -2$. **a** Normalized intensity $|E_{x,y}|^2$ and phase $\phi_{x,y}$ of electric field components $E_{x,y}$, normalized Stokes parameters $S_{1,2,3}$, ellipticity ε, as well as phase Φ_{12} with singularities marked red and insets visualizing the field twist for different propagation distances z. **b** Evolution of singularities in 3d space with z-slices of $\Phi_{12}(x, y, z)$. The chosen area is highlighted by a white dashed box in **a**, $\Phi_{12}(z = 0\,\mu\text{m})$. [Adapted with permission from Ref. [14]; © 2018 WILEY-VCH Verlag GmbH & Co. KGaA, Weinheim]

are discretized[6] according to $\varphi \in \{0, \frac{1\cdot 2\pi}{256}, \frac{2\cdot 2\pi}{256}, \frac{3\cdot 2\pi}{256}, ..., 2\pi\}$. These minor perturbations would already cause an individual non-generic singularity to unfold, i.e., a V-point or higher-order phase vortex in a purely polarization or phase structured field would split into first-order C-points or phase vortices upon propagation, respectively. Now these slight perturbations are added to the combination of a non-generic phase and polarization singularity, so that the interaction of unfolding properties can be examined. Note that, as proven in Ref. [14], experimentally, the considered situation in a perturbed system can be easily realized by a DPMS, which naturally includes the additional global phase variation (cf. Sect. 2.4.3). Here, the pixelated SLM as key component includes the intended spatial and phase value discretization as minor perturbations.

As illustrative example, a spider-shaped vector field, hence, a vector field with $\sigma_{12} = +4$ with additional higher-order phase vortex of charge $\ell = -\sigma_{12}/2 = -2$ is investigated [14]. Numerical results are presented in Fig. 3.6. Here, in Fig. 3.6a the electric field distributions and its polarization are depicted within three different planes: Enclosed by a dashed box, the intensity of the x- and y-component $|E_{x,y}|^2$,

[6]Consider that these discretizations perfectly mirror an according experimental realization of this investigation in which an SLM, or, more precisely, a DPMS would be applied, as demonstrated in Ref. [14].

the respective phase $\phi_{x,y}$, normalized Stokes parameters $S_{1,2,3}$, the states' ellipticity ε as well as the phase Φ_{12} of the light field within the 0-plane ($z = 0\,\mu$m) are shown. Thereby the 0-plane would correspond to the SLM plane in the respective experiment.[7] Additionally, the distributions of respective parameters are presented in a plane before and behind the 0-plane ($z = \{-950,\ 950\}\,\mu$m).

In the 0-plane, the light field reveals the intended distributions. Namely, Φ_{12} represents a phase vortex of index $\sigma_{12} = 4$, exhibiting the embedded non-generic V-point in a spider-shaped vector field (cf. gray flow lines). The polarization states' linearity is proven by $S_3,\ \varepsilon = 0$, whereas $S_{1,2}$ show spatially varying structures. The additional phase vortex can be identified within $\phi_{x,y}$: Typically, for a pure vector mode these distributions only include areas of discrete phase values 0 and π, as visible in Eq. (3.6) for $\ell = 0$. In contrast, here the sum of these areas plus a phase vortex of charge $\ell = -2$ is observable.

If the 0-plane is left, the unstable V-point instantly splits into first-order singularities, as visible in Φ_{12}. The number N of singularities is given by the index $\sigma_{12} = N$ of the V-point. Further, these singularities are located in a circular area of elliptically polarized states (cf. S_3, ε), whose size is increasing with z-distance. Hence, these singular points represents generic C-point singularities. Interestingly, these singularities are located equally distant on a circle around the optical axis. Moreover, upon propagation these singularities as well as the whole light field twist left- or right-handed for negative or positive values of z, respectively. Note that, for a negative index sign of σ_{12} (web configuration), this twist is right-/left-handed before/behind the 0-plane [14]. The twisting behavior is especially visible in Φ_{12} (see insets), but also in Stokes parameters $S_{1,2}$, the intensity and phase of $E_{x,y}$. To understand this phenomena in more detail, the phases $\phi_{x,y}$ need to be studied, as they reveal the effect of the included phase vortex. Here, $\phi_{x,y}$ clearly exhibit a splitting of additional higher-order, thus, unstable phase vortex originating from the minor perturbations based on discretizations. In combination with the areas of typically discrete phase values, the occurrence of multiple phase singularities is caused at the initial lines (see 0-plane) of phase dislocations of π. These singularities are also placed on circles and twist around the optical axis with propagation. This behavior obviously mirrors the typical unfolding properties of higher-order phase vortices and is transferred to the V-point splitting into C-points, twisting around the optical axis. Hence, besides observing the singularity conversion from non-generic V- to generic C-points, the sophisticated transfer of unfolding properties from the additional phase vortex to the splitting V-point is evinced [14] by the customized combination of non-generic scalar and vectorial singularity.

This unfolding evolution is highlighted in Fig. 3.6b in a 3d plot of singularities (red dots) and z-slices of Φ_{12}. Here, a small area of Fig. 3.6a (see white dashed box in Φ_{12} for $z = 0\,\mu$ m) is evaluated in 3d space. Obviously, point singularities in 2d form line singularities in 3d. This plot particularly emphasizes the instability of the V-point, as the repulsion of splitting singularities increases with decreasing

[7]Sizes in Figs. 3.6-3.8 (see scale bars) are scaled according to the potential experiment with SLM pixels of $8 \times 8\,\mu\text{m}^2$ in size.

distance to the 0-plane. This instant splitting of singularities, diverging with distance z, in the near-field of the tailored beam resembles an explosion, reasoning the term "singularity explosion" [14] for this phenomenon.

3.2.1.2 Hybrid Vector Field with Mixed Phase Vortices

Besides affecting the splitting of a single V-point by the addition of an unstable phase vortex in a slightly perturbed system, the interaction of propagation dynamics of multiple non-generic V-points and phase singularities can be observed within tailored light fields. Therefore, a hybrid vector field embedding both, a positively and negatively indexed V-point, is combined with a mixed vortex configuration. More precisely, a light fields following Eq. (3.6) with $\ell = -\sigma_{12}/2$ but with

$$\begin{aligned} \sigma_{12} &= +8 \quad \text{for} \quad \varphi = [3\pi/2,\ \pi/2[\,, \\ \sigma_{12} &= -8 \quad \text{for} \quad \varphi = [\pi/2, 3\pi/2[\end{aligned} \tag{3.7}$$

is exemplarily created (polar angle $\varphi \in [0,\ 2\pi]$). Consider that, again, in addition to the numerical evaluation, this field can be easily realized by the DPMS (cf. Sect. 2.4.3).

In this case, a hybrid vector field with two flower petals and four web sectors is formed as visualized in Fig. 3.7a (red: flow lines) with the respective electric field properties illustrated in Fig. 3.7b. Here, the petals see a negative azimuthal phase gradient whereas the web sectors are combined with a positive one, as visible in $\phi_{x,y}$. This manifests in the mirror symmetry of $\phi_{x,y}$ with the vertical line through the beam center (optical axis) as axis of symmetry. Following Eqs. (3.6) and (3.7), the tailored field embeds two very close V-points of index $\sigma_{12} = \pm 4$ as well as phase singularities of charge $\ell = \mp 2$. Note that both, V-points as well as phase vortices, of opposite charge can annihilate each other if they coincide [17]. Hence, depending on the distance of oppositely charged/indexed singularities, there are no or two singular points. However, the total index or charge for vectorial and scalar singularities is constant with $\sigma_{12}^{\text{tot}} = 0$ and $\ell^{\text{tot}} = 0$, respectively.

Figure 3.7c, d present the numerical results for the propagation dynamics. Here, the former includes the Stokes parameters $S_{1,2,3}$, the states' ellipticity ε as well as the phase Φ_{12} with marked polarization singularities (red dots/red circles filled white: positively/negatively indexed singularities) for three different planes $z = \{-950,\ 0,\ 950\}$ µm. The latter show the respective 3d plot of singularities propagating in 3d space with z-slices of Φ_{12}, whereby the chosen area is marked by a white box in Φ_{12} for $z = 0$ µm of Fig. 3.7c. Crucially, even though singularities may annihilate in the 0-plane, both V-points unfold upon propagation into generic first-order C-points (cf. Φ_{12}, S_3, ε in Fig. 3.7c). Following singularity index conservation, four positively and four negatively indexed C-points (/C-lines in 3d space) are observed. Also in this case the splitting is initiated by unfolding phase vortices in $\phi_{x,y}$ affected by the minor perturbations. Beyond, the light field twisting inherited from splitting

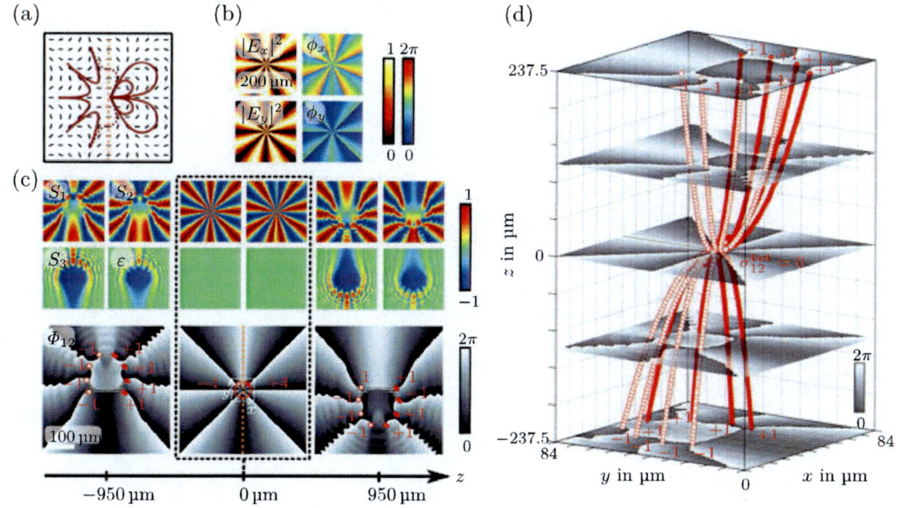

Fig. 3.7 Affecting the propagation dynamics by joining multiple phase and polarization singularities in a hybrid vector field. **a** Polarization structure of hybrid configuration (red: flow lines; orange: border of positive/negative index σ_{12}) with **b** normalized intensity $|E_{x,y}|^2$ and phase $\phi_{x,y}$ of electric field components $E_{x,y}$. **c** Normalized Stokes parameters $S_{1,2,3}$ and ellipticity ε as well as phase Φ_{12} with positively or negatively indexed singularities marked by a red dot or red circle filled white, respectively. **d** 3d plot of singuarity propagation in 3d space with z-splices of $\Phi_{12}(x, y, z)$. The chosen area is highlighted by a white dashed box in **a**, $\Phi_{12}(z = 0\,\mu\text{m})$

phase vortices is observable in the Stokes parameters $S_{1,2}$ as well as Φ_{12} in Fig. 3.7c and d. Interestingly, C-points originating from negatively or positively indexed V-points twist in opposite directions before as well as behind the 0-plane, leaving the impression of all singularities moving collectively in $+y$- and $-y$-direction for negative and positive z-propagation, respectively. This unique behavior is particularly visible in Fig. 3.7d and highlights the adaptability of singularity propagation dynamics by the customization of the light field.

3.2.2 *Converting Point and Line Singularities in Ellipse Fields*

Besides appearing in linearly polarized vector fields, non-generic V-points may also occur in ellipse fields, as presented in Sect. 2.3.2. This V-point corresponds to the intersection of generic L- and non-generic C-lines. The respective ellipse field, purely varying in polarization, can be defined as

$$\vec{E} = \frac{1}{\sqrt{2}} \left[\mathrm{e}^{-\mathrm{i}\frac{\sigma_{23}}{2}\cdot(\varphi+\alpha_1)},\ \mathrm{e}^{+\mathrm{i}\frac{\sigma_{23}}{2}\cdot(\varphi+\alpha_2)} \right]^T \tag{3.8}$$

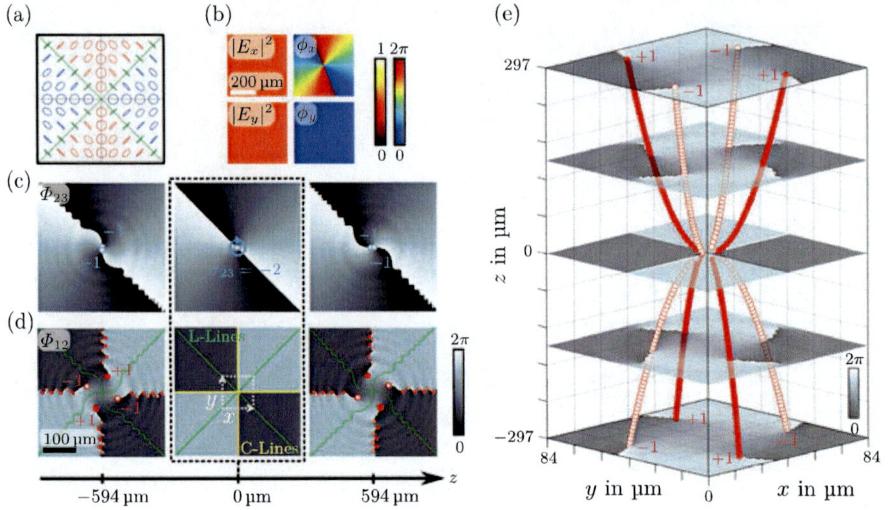

Fig. 3.8 Guiding point and line singularities in an ellipse field. **a** Polarization configuration of Poincaré vortex structure (blue/red/green: left-/right-handed elliptical/linear polarization) with green L- and gray C-lines in the 2d 0-plane and **b** respective electric field properties. **c, d** Phases $\Phi_{23,12}$ upon propagation with point singularities marked by colored circles filled white (negative index) or dot (positive index), L-lines in green and C-lines in yellow. **e** 3d plot of propagating Φ_{12} singularities (C-point quadrupole) close to the optical axis with Φ_{12}-slices. [Adapted with permission from Ref. [14]; © 2018 WILEY-VCH Verlag GmbH & Co. KGaA, Weinheim]

with $\alpha_{1,2}$ defining a phase off-set between x- and y-components and σ_{23} representing the index of embedded V-point in the so-called "Poincaré vortex". Again, a phase vortex is added to this configuration as a tailored perturbation, so that the studied light field is calculated from

$$\vec{E} = \frac{1}{\sqrt{2}} \left[\mathrm{e}^{-\mathrm{i}\frac{\sigma_{23}}{2}\cdot(\varphi+\alpha_1)},\ \mathrm{e}^{+\mathrm{i}\frac{\sigma_{23}}{2}\cdot(\varphi+\alpha_2)} \right]^T \cdot \mathrm{e}^{\mathrm{i}\ell\varphi}\,. \tag{3.9}$$

As an example, $\alpha_1 = 3\pi/4$, $\alpha_2 = \pi/4$, $\sigma_{23} = -2$ and $\ell = -\sigma_{23}/2$ is chosen. The respective light field is visualized in Fig. 3.8a, b. In Fig. 3.8a the polarization distribution is depicted by small polarization ellipses (red/blue/green: right-/left-handed elliptical/linear polarization) with green (gray) lines marking L-lines (C-lines). The intensity $|E_{x,y}|^2$ and phase $\phi_{x,y}$ of the electric field components $E_{x,y}$ are presented in Fig. 3.8b. Similar to vector fields, this tailored field is realizable by the DPMS with $\beta_1 = -45°$, $\beta_2 = 90°$ and $\Delta\Phi_{\mathrm{SLM}} = \mod[-2\varphi + \pi/4,\ 2\pi]$ [14].

3.2.2.1 Unfolding Point Singularity

By the addition of the vortex structure to the polarization modulated field, a double-charged vortex is created in ϕ_x, hence, it includes a non-generic higher-order phase singularity. Considering the minor perturbations realized by discretization, this singularity is expected to unfold and transfer its unfolding properties to the V-point, embedded in the ellipse field [14]. By studying the phase Φ_{23}, used for identification of V-points in ellipse fields, this behavior can be verified, as illustrated by numerical results in Fig. 3.8c. Here, Φ_{23} is examined in three different planes with $z = \{-594, 0, 594\}$ μm. In the 0-plane Φ_{23} reveals the embedded V-point of index $\sigma_{23} = -2$. Due to the effect of the unfolding additional phase vortex in the perturbed system, this non-generic V-point unfolds upon propagation with a right- and left-handed twist in the light field before and behind the 0-plane, respectively. The arising first-order singular points ($\sigma_{23} = -1$), forming line singularities in 3d space, correspond to stationary points of vertically oriented linear polarization [14]. Hence, also in the ellipse field the transfer of the unfolding properties of a higher-order scalar singularity—here in ϕ_x—to the polarization singularity is evinced.

3.2.2.2 Propagation Dynamics of Line Singularities

Besides point singularities evolving in 3d space, the propagation dynamics of generic as well as non-generic line singularities can be investigated within this light field configuration. Therefore, the phase Φ_{12} is studied, as shown in Fig. 3.8d. In the 0-plane non-generic C- (yellow) as well as generic L-lines (green) are observed in Φ_{12},[8] which solely contains areas of constant phase $\pi/2$ and $3\pi/2$. Upon propagation, these areas break up and merge into a continuous flow of phase values. Crucially, this merging procedure has no effect on the stability of generic L-lines (in 3d: L-surfaces), which only bend according to the light field twist. In contrast, the change in Φ_{12} goes along with the splitting of non-generic C-lines into multiple generic first-order C-points [14], proving the instability of C-lines/surfaces in 2d/3d space. In 2d, located on circular rings of different radii around the optical axis, C-points form quadrupoles with two positively and two negatively indexed singular points ($\sigma_{12} = \pm 1$). Hence, the total singularity index $\sigma_{12}^{\text{tot}} = 0$ is conserved within the light field. Similar to the behavior of splitting V-points, C-points also twist with the whole light field upon propagation—right-/left-handed before/after the 0-plane. Thus, also for the occurrence of C-points an adoption of the behavior of unfolding scalar phase singularities is demonstrated.

Figure 3.8e emphasizes the propagation dynamics of arising C-point singularities. Here, the evolution of the most central quadrupole is shown with Φ_{12}-slices (chosen area: white dashed box in Fig. 3.8e), revealing the formation of generic C-lines in 3d space. The C-lines diverge with distance from the 0-plane, whereby no point singularities are found in this 2d 0-plane. In total, starting with negative z-values, one observes a quadrupole of C-points in 2d space converging with increasing propagation distance, annihilating each other in the 0-plane and emerging again, diverging

---

[8] As outlined in Ref. [14], C- or L-lines in a transverse plane can be identified as phase dislocations of π in Φ_{12} or phase values of $\{0, \pi, 2\pi\}$ in Φ_{23}, respectively.

with z-distance, if the 0-plane is left. Hence, the annihilation as well as creation of C-points is evinced.

3.2.3 Conclusion

In this section the lack of knowledge about combined scalar and vectorial singularities in 3d space has been tackled. By numerically combining polarization and phase structuring in a slightly perturbed system, the control of polarization singularity propagation dynamics was demonstrated. Therefore, non-generic scalar and vectorial singularities were joint in a tailored vector and ellipse field [14], by which, in both cases, the transfer of unfolding properties from one singularity class to another was proven. Beyond, a hybrid vector field with mixed phase vortices gave new insights into the propagation dynamics of point singularities of the same and different classes in very close vicinitiy. It was demonstrated that even though non-generic vectorial/scalar singularities of opposite topological index/charge may annihilate each other in a specific plane, the splitting of each singularity into its generic first-order counterparts can be created by structured light techniques. Additionally, the genericity of L- and C-line singularities (2d space) was studied under minor perturbations [14], proving the stability of the former and instability of the latter. For C-lines, the creation and annihilation of multiple C-point quadrupoles, located on circular rings and twisting around the optical axis, was evinced and explained by numerical evaluations. All studied and controllable phenomena were observed in the near-field of customized light fields, highlighting the instability of non-generic singularities under perturbations.

3.3 Non-diffracting and Self-imaging Vectorial Light Fields

In previous sections the realization of complex vectorial modes as well as the customization of singularity propagation dynamics in tailored light fields were demonstrated. In the case of the former, IVMs were proposed as a novel overall class of self-similar vectorial light fields, embedding conserved networks of singular points. In the latter case, tailored fields shaped in phase and polarization were considered for the study of the interactions of singularity propagation properties and, thus, the control of respective dynamics. In both cases, the according vectorial light field was structured in 2d space and subsequently observed upon propagation in 3d. In contrast, here, the innovative 3d customization of singular light fields even including complex SNs is demonstrated. For this purpose, scalar *non-diffracting* as well as *self-imaging* light fields will be considered as basis for the vectorial field customization revealing unique 3d propagation properties.

As indicated by its name, upon propagation scalar non-diffracting or propagation invariant light fields do not change in their transverse appearance including their amplitude and phase. Naturally, the homogenous polarization of these fields does not vary neither. These fields are considered as tailored or structured in 3d space since, in contrast to typical diffracting fields enlarging their transverse diameter upon propagation, non-diffracting beams are customized to keep their transverse appearance including its diameter. Note that although theoretically the distance along which the light field is non-diffracting is infinite, experimentally this distance has a limited length (see Sect. 3.3.1). A unique property assigned to these fields is the self-reconstruction—"self-healing"—of its properties if the beam is obstructed by an obstacle. Due to their valuable propagation-invariance as well as the self-healing property, these diffraction-free beams have already found applications, e.g., in imaging [39–41], laser material processing [42], or optical trapping [43–46]. Interestingly, the property of self-healing is not limited to diffraction-free fields, but can also be found in other structures as in caustic or self-similar fields, namely Airy [47], Pearcey [48], LG [49, 50], or even standard fundamental Gaussian beams [51]. Beyond, this property is also found in so-called self-imaging light fields [52, 53]. These fields have the interesting characteristic of periodically replicating their transverse distribution upon propagation [52], which prove beneficial for, e.g., the enhancement of electron microscopy resolution [54] or the realization of integrated optics [52, 55, 56].

Although these 3d customized fields have been studied intensively in the scalar case, its vectorial analogon, i.e., a realization of these fields additionally embedding an inhomogenous longitudinal and/or transverse distribution in polarization tailored in 3d space, only received little attention so far. Recent studies mainly focused on the formation of non-diffracting polarization configurations (transversely inhomogeneous), whereby their customizability is clearly limited so far. For instance non-diffracting Mathieu-Poincaré beams were introduced [57], realized by incoherent superposition of scalar Mathieu beams, hence, similarly to CVBs and IVMs, but adopting the non-diffracting properties of Mathieu fields. By the same principle vector Bessel modes [15, 58–61] can be formed, mathematically described as the combination of orthogonally polarized higher-order scalar Bessel beams, as outlined in the following section. Interestingly, these non-diffracting vector modes pave the way to novel, more robust techniques in, e.g., quantum key distribution utilizing the non-separability within these modes of light (cf. Chap. 5). However, as these fields are based on Bessel as well as Mathieu beams, the diversity of respective vectorial fields is limited by the mode indices of these scalar fields.

In order to inhomogeneously structure polarization in 3d space, i.e., including longitudinal modulation, first attemps were performed by different authors, mainly based on scalar Bessel fields which were additionally shaped by, e.g., holographic means, a retro-reflector and/or wave plates [62–64]. By these approaches the polarization was purely sculpted in the longitudinal direction or both, longitudinally and transversely. Due to the Bessel basis vectorial light fields are again limited in their diversity and can only include single on-axis singularities.

Overcoming the restrictions of known 3d structured vectorial fields, in this section, vectorial discrete non-diffracting as well as vectorial self-imaging modes are presented. As a fundamental basis, vector Bessel modes are introduced firstly, experimentally realized by a combination of SLM and q-plate and investigated in its self-healing properties all-digitally [15]. Following, vectorial discrete non-diffracting fields with an impressive number of DoFs, embedding complex SNs, are shown. Based on this study, the novel class of vectorial self-imaging beams is outlined. They are proven and studied experimentally, revealing the complexity of respective amplitude, phase, polarization and singularity structures in 3d space. Crucially, both, discrete non-diffracting as well as self-imaging fields including their networks of polarization singularities adopt the propagation properties of their scalar ancestor.

3.3.1 *Self-healing Vector Bessel Modes*

3.3.1.1 Bessel-Gaussian Beams

Arguably, the best-known non-diffracting light fields are Bessel modes, firstly introduced by J. Durnin in 1987 [65, 66]. As, theoretically, Bessel modes would need an infinite amount of energy to stay unchanged for infinite propagation distances and, additionally, to be infinitly expanded in the transverse plane, Bessel-Gaussian (BG) beams were introduced as a valid approximation over finite propagation distances [67]. Within this finite distance, BG fields have the same property of non-diffraction and the ability of self-reconstruction in amplitude and phase as Bessel modes [44, 68]. BG beams are described by

$$E_\ell^{\mathrm{BG}}(r, \varphi, z) = \sqrt{\frac{2}{\pi}} J_\ell\left(\frac{z_R k_r r}{z_R - \mathrm{i}z}\right) \exp(\mathrm{i}\ell\varphi - \mathrm{i}k_z z) \cdot \exp\left(\frac{\mathrm{i}k_r^2 z w_0 - 2kr^2}{4(z_R - \mathrm{i}z)}\right) \tag{3.10}$$

with polar coordinates (r, φ, z), ℓ being the azimuthal index (topological charge), and k_r and k_z representing the radial and longitudinal wave numbers, respectively. While $J_\ell(\cdot)$ defines the Bessel function, the Gaussian information is found in the last factor with the initial beam waist w_0 of the Gaussian profile, limiting the transverse beam expansion. This light field can be seen as a superposition of plane waves with wave vectors lying on a cone described by the angle $\Theta = k_r/k$ (k: wave number) [67]. This is exemplified in Fig. 3.9a. Obviously, a rhombus-shaped region of length $z_{\max}$ ("non-diffracting distance") is formed, in which the light field is non-diffracting. For small angles $\Theta \approx \sin\Theta$, the length of the non-diffracting region is calculated from $z_{\max} = 2\pi w_0/\lambda k_r$ [69]. The center of the region is positioned at z_0. If an obstruction of radius R is placed within $z_{\max}$, a shadow region of length $z_{\min} \approx R/\Theta \approx \frac{2\pi R}{k_r \lambda}$ is formed [70]. Due to plane waves passing the included obstruction, the light field starts to recover after the self-healing distance $z_{\min}$ [44, 68] and is fully reconstructed

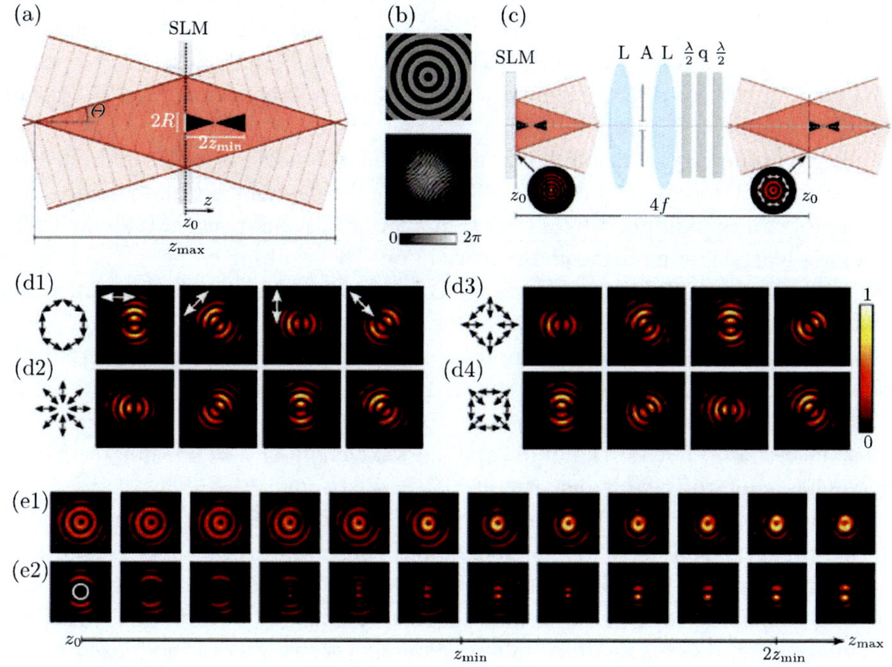

Fig. 3.9 Realization and investigation of self-healing vector Bessel-Gaussian (vBG) modes. **a** Concept of obstructed scalar BG beam, **b** binary Bessel function (top) and phase-only hologram (bottom) for the formation of a BG mode, and **c** sketch of experimental system for the creation of vBG beams (L: lens with focal distance f, A: aperture, $\frac{\lambda}{2}$: half-wave plate, q: q-plate). **d** Experimental examples of vBG modes analyzed with respect to their polarization (black arrows) by a rotating polarizer (orientation: white arrows). **e** Experimental study of 3d propagation properties of (e1) an azimuthally polarized vBG beam and (e2) the same beam but obstructed by an absorbing obstacle (white circle) of radius $R = 200\,\mu\text{m}$ and additionally filtered by a polarizer (horizontally oriented). [Adapted with permission from Ref. [15]]

at $2z_{\text{min}}$, as illustrated in Fig. 3.9a (first/second black cone: obstacle shadow/recovery region).[9]

For the experimental realization of BG fields a SLM represent a well-suited tool, as it allows for an on-demand dynamic modulation of the scalar field properties by computer generated holograms. For this purpose, the binary Bessel function is chosen as phase-only hologram with the transmission function

$$T(r, \varphi) = \text{sign}\{J_\ell(k_r r)\} \exp(\text{i}\ell\varphi) \tag{3.11}$$

---

[9]Note that the field is considered as *fully* recovered at $2z_{\text{min}}$, since, at this point, the light field completely reconstructed the specific information cut by the obstacle (in Fig. 3.9a: around the optical axis at $z = z_0$). Thus, at $2z_{\text{min}}$ the transverse plane of the field shows an undisturbed field distribution in the area of radius R around the optical axis. At z_{min}, the field starts to reconstruct this area.

including the sign function sign$\{\cdot\}$ [71, 72]. An exemplary phase distribution of this function is shown in Fig. 3.9b (top). The according hologram causes the respective BG mode to appear directly behind the SLM with z_0 representing the SLM plane (cf. Fig. 3.9a).

For an exemplary realization of a fundamental BG mode, $k_r = 18\,\text{rad mm}^{-1}$ and $\ell = 0$ are set (Holoeye Pluto SLM). To achieve the demanded Gaussian envelope, the hologram is multiplied by a Gaussian aperture function ($w_0 = 0.89\,\text{mm}$) after adding a blazed grating to the phase distribution. The resulting hologram is shown in Fig. 3.9b (bottom). By this approach, the BG mode is realized in the first diffraction order of the blazed grating following the principle of joint amplitude and phase modulation (cf. Sect. 2.4.1). By applying a $4f$-imaging system, as conceptionally shown in Fig. 3.9c, the first order can be filtered. The intensity distribution of the created BG mode detected in the z_0-plane, i.e., image plane of the SLM, is visualized as an inset in Fig. 3.9c (left). For the applied wavelength of $\lambda = 633\,\text{nm}$, the formed mode has a non-diffracting distance of $z_{\max} = 49.16\,\text{cm}$.

Besides applying it for the generation of BG modes, the SLM can also be used for the all-digital and on-demand implementation of absorbing as well as phase obstacles within the z_0-plane [15], as indicated in Fig. 3.9c. This enables the investigation of the self-healing field properties with tailored obstructions. For absorbing obstructions, a circular central cut is included in the hologram, so that there is no blazed grating within the area of obstruction. Phase obstacles can be created by adding the chosen phase object to the hologram. By imaging the SLM and, thus, the BG mode, the embedded obstruction is also projected into the plane of analysis (cf. Fig. 3.9c).

3.3.1.2 Creating Vector Bessel-Gaussian Modes

In order to go from scalar to vector BG (vBG) modes, different techniques can be applied, such as interferometric or holographic approaches (cf. Sect. 2.4) as well as optical elements designed for this specific purpose, namely, a so-called q-plate. This device consists of a thin layer of liquid crystals between two glass plates, whose fast axis are arranged in a singular pattern with topological charge q [73]. By applying suitable voltage, the retardation of this plate can be tuned to be optimally π [74], so that the q-plate acts as a standard HWP with a spatially inhomogeneous orientation of its fast axis. In this case, the device is able to change the OAM or, more precisely, its topological charge ℓ, of a circularly polarized beam by $\pm 2q$ depending on the polarization handedness. Considered in the Jones formalism, the q-plate is defined by the matrix

$$Q(q, \varphi) = \begin{pmatrix} \cos(2q\varphi) & \sin(2q\varphi) \\ \sin(2q\varphi) & -\cos(2q\varphi) \end{pmatrix} \tag{3.12}$$

with linear polarization basis [75]. Considering, for instance, a horizontally linear input polarization, the q-plate in combination with chosen wave plates can realize various vectorial modes [8, 15, 75].

In the case at hand, a q-plate (q) combined with one or enclosed by two HWPs ($\frac{\lambda}{2}$) is used for the addition of polarization structuring to considered BG modes. Working in the paraxial regime, this set of optical elements can be positioned arbitrarily within the beam path. Here, the set is located directly before the non-diffracting region of the linearly polarized (horizontal) fundamental BG beam [15], as illustrated in Fig. 3.9c. Note that it could also be positioned within z_{max} [75] or even in its far-field [61]. By the applied technique, depending on the orientation $\beta_{1,2}$ of the first and/or second HWP, vBG modes imparting a transversely inhomogenous distribution in polarization are formed, as indicated by the inset of an azimuthally polarized vBG mode. Figure 3.9d shows four different examples of created vBG fields (without obstacle) studied at the z_0-plane with a polarizer in front of a camera (normalized intensity images). The polarizer's orientation is indicated by white arrows (passing polarization), revealing the light fields' polarization, which is also depicted by black arrows on the left. Originating from the polarization configurations, all shown examples theoretically embed a V-point singularity ($\sigma_{12} = \pm 2$) of undefined polarization on-axis, i.e., in its dark center. However, consider that these non-generic singularities are probably unfolded into generic C-points due to experimental perturbation. The respective structures were realized by settings of (d1) $\beta_1 = \pi/4$ (no second HWP), (d2) $\beta_1 = 0$ (no second HWP), (d3) $\beta_1 = \pi/4$, $\beta_2 = 0$, and (d4) $\beta_1 = 0$, $\beta_2 = 0$.

3.3.1.3 Self-healing Vectorial Modes

Equivalent to scalar BG modes, vBG beam impart the same 3d propagation properties, namely, they are non-diffracting as well as self-healing if being obstructed. This is due to the fact that vBG modes embody the incoherent superposition of two higher-order scalar Bessel modes. More precisely, in circular polarization basis ($\vec{e}_{R,L}$: right-, left-handed circular polarization) vBG modes, as realized above, can be written as

$$\vec{E}^{\text{vBG}} = \frac{1}{\sqrt{2}} \left[\vec{e}_R \cdot E_{\ell}^{\text{BG}} \cdot \mathrm{e}^{\mathrm{i}\alpha} + \vec{e}_L \cdot E_{-\ell}^{\text{BG}} \cdot \mathrm{e}^{-\mathrm{i}\alpha} \right]. \tag{3.13}$$

with relative phase difference α and $\ell \neq 0$ for achieving the vectorial character. Thus, vBG beams inherit the respective properties of imparted scalar BG modes, similarly to CVBs being composed of higher-order LG beams and inheriting their self-similar property, i.e., structural stability (cf. Sect. 3.1.1). Crucially, respective properties are also expected for included singularities, i.e., V-/C-points (2d space), as the transverse polarization structure is propagation invariant or self-heals if being obstructed. While in an idealized case, V-points would show respective non-diffracting/self-healing behavior, in a non-idealized case as in the experiment, unfolded non-generic V-points, i.e., C-points as its generic counterparts (C-point in 2d/C-lines in 3d) would be observed.

For an experimental proof of inherited properties, an azimuthally polarized vBG mode (cf. Fig. 3.9d1) is created as described above and studied upon propagation. For this purpose, first, the camera is moved longitudinally through one half of the non-diffracting rhombus-shaped region of the 3d structured field, i.e., from the z_0-plane to $z_{\max}$, visualizing the intensity in different (x, y)-planes in 3d space. By this technique the non-diffracting property within this region is evinced [15], as visible by the results in Fig. 3.9e1. Consider that upon propagation in $+z$-direction the outer rings of the vBG mode are more and more disappearing, which reflects the rhombus shape of the non-diffracting region.

Second, for the investigation of the vBG modes self-healing property [15], an absorbing obstacle with $R = 200\,\mu\text{m}$ is included digitally in the SLM plane. As the SLM initially shapes a fundamental BG mode with $\ell = 0$ and the embedded obstacle, and is subsequently projected through the q- and wave plate set into its image plane, the created vBG mode also imparts the tailored obstacle in its z_0-plane (cf. Fig. 3.9c). Performing the same measurement as before but with an additional polarizer (horizontally aligned) in front of the camera, moving simultaneously, the self-healing of the amplitude, phase as well as polarization and, indirectly, associated singularities is evaluated upon propagation. Results are shown in Fig. 3.9e2. In general, the typical petal structure, occurring if a polarizer is included in a vector field, is observed. Obviously, the imparted digital object casts a shadow upon propagation, whereby the light field starts to recover the region of obstruction when $z_{\min}$ is reached. For the programmed obstacle, a self-healing distance of $z_{\min} = 11.02\,\text{cm}$ is calculated, which is confirmed by the experiment. Shortly before the non-diffracting region ends, the complete self-reconstruction of the light field and its DoFs (amplitude, phase and polarization) is found at $2z_{\min}$.

3.3.2 *Discrete Non-diffracting Vector Fields*

As outlined above, scalar BG modes of light can be considered as a superposition or interference of plane waves with wave vectors lying on a cone with cone angle Θ. In the far-field, these waves form an (ideally) infinitesimal thin circular ring representing the wave vectors $\vec{k}_j$ in optical Fourier or k-space. Naturally, even two plane waves with k-vectors $\vec{k}_{1,2}$ enclosing an angle of 2Θ would result in a non-diffracting longitudinal interference structure representing the simplest form of a so-called *discrete non-diffracting (dnd) field*. The notation "discrete" originates from the fact that, in k-space, the respective plane waves represent two discrete points (spatial frequency components) positioned on a joint ring. This knowledge allows for the realization of a great variety of discrete non-diffracting light fields: Within the Fourier/k-space multiple plane waves on a joint ring can be chosen which interfer in its respective real space. Crucially, not only the number of plane waves on a joint ring can be varied, but also the phase relation between them as well as the relative angular position of waves on the ring [76]. These DoFs explain the diversity of discrete non-diffracting fields.

Mathematically, in real space these scalar fields are defined as [76]

$$E_{\mathrm{dnd}}(x, y, z) = E_0 \sum_{j=1}^{N} \exp\left\{\mathrm{i}\left(k_{x,j}x + k_{y,j}y + k_z z + \phi_j\right)\right\} \tag{3.14}$$

with amplitude E_0, the wave vector $\vec{k}_j = [k_{x,j}, k_{y,j}, k_z]^T$, and relative phase ϕ_j per plane wave $j \in \{1, ..., N\}$. Note that if the phase is chosen as $\phi_j = 2\pi\ell j/N$ with topological charge ℓ, following singularity index conservation rule, this topological charge is found in k- as well as real space [76]. If the number of plane waves is infinite, so that they form a complete ring in k-space, the BG mode is formed. In this case, $|\ell| \geq 1$ results in higher-order BG modes carrying OAM.

Here, an approach is proposed for the realization of vectorial discrete non-diffracting beams, enabled by the inclusion of polarization $\vec{e}_j$ as an additional DoF within the beam class. Therefore, above's equation is transformed into

$$\vec{E}_{\mathrm{dnd}}(x, y, z) = E_0 \sum_{j=1}^{N} \vec{e}_j \cdot \exp\left\{\mathrm{i}\left(k_{x,j}x + k_{y,j}y + k_z z + \phi_j\right)\right\}. \tag{3.15}$$

Thereby, in the case of totally incoherent superposition with no joint polarization components of plane waves, pure polarization interference [77] excluding complex intensity configurations (s_0) will be observed. To create an amplitude, phase and polarization structured field in real space, at least two plane waves/frequency components in Fourier space need to have joint polarization components, so that the superposition in real space is partially coherent and interference is possible. By this technique, the resulting field inherits non-diffracting and self-healing properties of scalar discrete non-diffracting beams.

3.3.2.1 Experimental Realization

For the experimental implementation of the above introduced novel beam class, a far-field construction method [78] combined with an APP modulation system (cf. Sect. 2.4.3) is applied. A sketc.h of the respective experimental system is illustrated in Fig. 3.10a. Here, a SLM is used in split-screen mode, whereby the first half is responsible for shaping the amplitude ($A^{\mathcal{F}}_{\mathrm{dnd}} = |\mathcal{F}[\vec{E}_{\mathrm{dnd}}]|$) and phase ($\phi^{\mathcal{F}}_{\mathrm{dnd}}$) of the vectorial non-diffracting beams in Fourier space, thus a Fourier hologram is applied with the DoFs encoded in a blazed grating (cf. Fig. 3.10b1). Due to experimental constraints, the frequency components representing points in k-space are approximated by circles with a diameter of 19 px within the amplitude modulation by the SLM. By this approach, a sufficient amount of light is reflected by the SLM. Further, the first SLM half is imaged onto the second by a $4f$-system consisting of two lenses (L) between which the first diffraction order is filtered by an aperture (A). The second SLM half, being part of a DPMS (QWP$_1$-SLM-QWP$_2$; $\beta_{1,2} = -\pi/4$ ($-45°$)), imprints the respective polarization ($\vec{e}_j$) onto each plane wave j, also acting in the far-field. Exemplary phase holograms for the first and second pass are

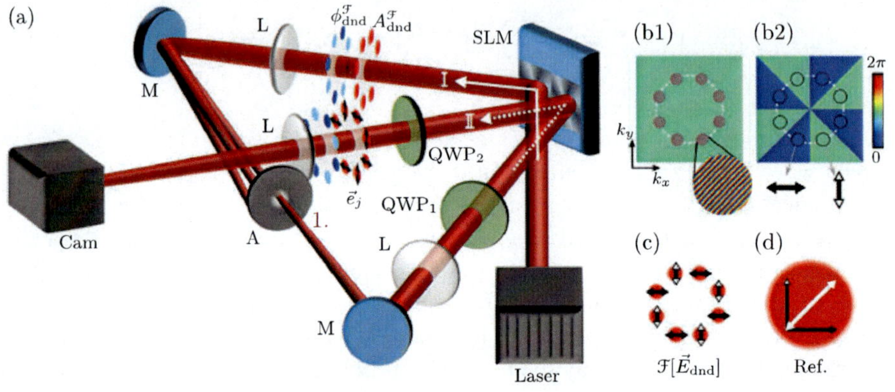

Fig. 3.10 Experimental realization of vectorial discrete non-diffracting light fields. **a** Experimental setup for APP modulation in k-space (SLM plane) being transformed into real space (camera plane). **b** Fourier holograms for (b1) amplitude and phase as well as (b2) polarization modulation. **c** Realized distribution $\mathcal{F}[\vec{E}_{\mathrm{dnd}}]$ in k-space being superimposed with **d** a plane wave of diagonally linear polarization as reference beam (Ref.) for phase measurement

presented in Fig. 3.10b1, b2, respectively. In the presented case, neighboring waves on a ring have orthogonal linear polarizations (horizontally and vertically linear). By Fourier transforming this k-space distribution by a lens, the according vectorial non-diffracting light field is created in real space,[10] where a camera is positioned. For analyzing the polarization distribution, a rotatable QWP and a polarizer can be added (cf. Sect. 2.4.2). If the study of the light field properties in k-space is desired, the last lens is exchanged for a set of two lenses, imaging the SLM onto the camera.

As illustrative examples, two non-diffracting light fields are formed by the above explained far-field construction method. First, $N = 8$ plane waves equally distant on a joint ring are supposed to be of the same amplitude as well as phase and sculpted in polarization as indicated in Fig. 3.10c. For this purpose, the hologram of the second pass (II) is given in (b2). The hologram for the first pass (I) is illustrated in (b1), whereby a correcting phase is considered, compensating for the DPMS created phase modulation of $\Delta\Phi_{\mathrm{SLM}}/2$ (cf. Sect. 2.4.3). As second example, the same distribution is chosen but the DPMS caused phase modulation is not compensated but used as additional phase variation in k-space (phase difference of $\pi/2$ between neighbors). The results are presented in Fig. 3.11a–c (1. example) and d–f (2. example). Here, the chosen DoF are visualized in Fig. 3.11a/d with the global phase being color encoded ($\in [0,\ 2\pi]$). The resulting intensity ($s_0 \in [0,\ 1]$) and polarization distribution ($S_{1,2,3,} \in [-1,\ 1]$) in real space (0-plane) is illustrated in Fig. 3.11c/f with simulations (top, based on Eq. (3.15)) and experimental results (bottom) gained by Stokes parameter measurement.

[10]Note that due to the approximation of plane waves as circles in k-space on the SLM, the transverse distribution in real space is slightly blurred as frequency components not corresponding to the light field are included.

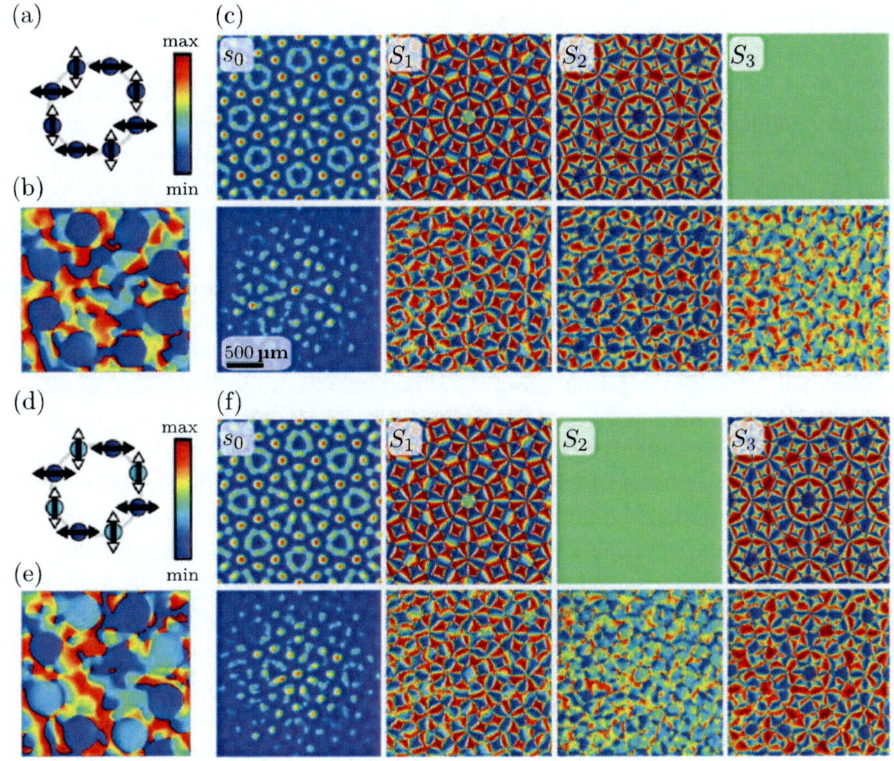

Fig. 3.11 Experimental examples of vectorial discrete non-diffracting light fields. **a/d** Sketch of configuration with chosen DoFs (phase: color encoded, $\in [0,\ 2\pi]$) and **b/e** measurement results of global phase ($\in [0,\ 2\pi]$) in k-space. **c/f** Numerical simulations (top) and experimental results (bottom) of respective Stokes parameters $s_0 \in [0,\ 1]$ and $S_{1,2,3} \in [-1,\ 1]$ in real space, $z = 0$

In both examples, experimental results perfectly mirror the calculated simulations. Obviously, complex structured vectorial light fields are formed, going far beyond basic polarization interference [77]. The intensity structures form an quasi-periodic pattern with an 8-fold rotational symmetry, originating from the eight plane waves being superimposed. Also the non-zero Stokes parameters reveal a structure of octagonal symmetry due to the number of frequency components. Note that these fields embed polarization singularities (more details discussed later on), more precisely, V-points in an idealized case (simulations), whose unfolding due to perturbations causes deviation between simulations and experimental results. Further, differences between theory and experiment especially visible in (Fig. 3.11c) $S_3 = 0$ or (Fig. 3.11f) $S_2 = 0$ are also due to inaccuracies of applied wave plates' retardation ($\pm\lambda/200$) and polarization sensitivity of optical components as metallic mirrors and lenses.

Even though the intensity only reveal minor changes if global phase modulation is included for the second example (Fig. 3.11d) in comparison to the first (Fig. 3.11a), the additional variation causes major changes in polarization: While the first example (Fig. 3.11c) represents a vector field with $S_3 = 0$, the phase variation in k-space causes the formation of an ellipse field with $S_3 \in [-1, 1]$ for the second example (Fig. 3.11f). However, the 8-fold rotational symmetry is kept—hence, by adapting the number of plane waves, the symmetry of distributions can be tailored. Further, comparing the first and second example reveals that the distributions initially observed for S_2 and S_3 (1. example) is now found for S_3 and S_2 (2. example), respectively, highlighting the adaptability of the transverse structure of the non-diffracting light field by the different DoFs in k-space.

Typically, if the polarization of a light field is structured, the additionally included global phase modulation is "invisible"—it cannot be measured by standard phase measurement techniques based on (coherent) interference [30, 79] or Stokes polarimetry [80]. In the former case, the superposition of the phase structured field with a reference beam of the same polarization under a small angle is demanded. For the latter, the phase structured field is combined on-axis with a beam of orthogonal polarization and resulting Stokes parameters are used to calculate relative phases. As the to be investigated light field's polarization is spatially inhomogeneous, these techniques are not applicable. Consequently, to prove which DoFs are tailored for the creation of above's discrete non-diffracting light fields, a smart method, based on off-axis interference, is proposed for the verification of applied phase variation.[11]

In the presented examples horizontal and vertical polarization are used for the spatial frequencies, whereby neighboring components in k-space shall have no phase shift (1. example) or a phase shift of $\pi/2$ to each other (2. example). For measuring this phase distribution, the SLM and, thus, the k-space is imaged onto the camera by two lenses. Now, this structure $\mathfrak{F}[\vec{E}_{\text{dnd}}]$ (Fig. 3.10c) is superimposed under a small angle with a reference beam (plane wave) of diagonal polarization, as indicated in Fig. 3.10d. The diagonal state embeds both, vertical as well as horizontal polarization, in equal portions and being in phase, whereby the horizontal/vertical part of the reference beam interferes with the horizontally/vertically polarized frequency components of $\mathfrak{F}[\vec{E}_{\text{dnd}}]$. By numerical Fourier transform of the resulting interference pattern, filtering the first order and inversely Fourier transforming it, a complex field is calculated. The phase of this field is equal to the global phase of the polarization structured field in k-space. Following this principle, for the first and second experimental example of a vectorial non-diffracting field, the respective phase modulations ($\in [0, 2\pi]$) in k-space are determined and presented in Fig. 3.11b, e, respectively. Obviously, the results verify the intended modulation, as for the first example no phase difference is detected between the frequency components, whereas a phase variation of $\pi/2$ is determined for spatial frequencies in the second example. Vice versa, the experimental results in comparison to the encoded modulation reveal the functionality of the proposed phase measurement method for the light field of inhomogenous polarization.

Note that, if multiple different states of polarization would be included in the analyzed beam, the polarization of the reference beam could be rotated and various interference patterns recorded. The respective measurement result combined with

[11] Amplitude and polarization can be verified easily by the camera and a polarizer.

before determined polarization configuration could be used to identify the global phase (position dependent) of the amplitude, phase and polarization, thus, fully-structured field.

3.3.2.2 Polarization Singularity Dynamics

Finally, the 3d propagation dynamics of the polarization structured non-diffracting field and, in particular, of its singularities are studied numerically. For this purpose, the light field of the first example (Fig. 3.11a–c) is analyzed with respect to embedded propagating singular points, hence, singular lines in 3d space. As the light field represents a vector field, the phase $\Phi_{12}(x, y, z)$ is calculated for a chosen propagation distance of $z \in [0, 150\,\text{mm}]$ with singularities being detected as intersections of zero-lines $Z_{\Re}$ (red, $S_1 = 0$) and $Z_{\Im}$ (blue, $S_2 = 0$) in transverse (x, y)-planes. Multiple singular points representing non-generic V-points (2d space) in the vector field, are found within the light field's transverse planes constituting complex arrangements within a SN. However, here, the analysis is exemplarily limited to the singularities closest to the optical axis, revealing the singularity propagation dynamics in vectorial discrete non-diffracting fields as representative examples.

The numerical results are presented in Fig. 3.12 with V-points indexed positively ($\sigma_{12} = 2$) or negatively ($\sigma_{12} = -2$) being marked by a red or blue (a) dot/(b) circle per analyzedplane, respectively. In Fig. 3.12a, the 3d plot of propagating singular points in $\Phi_{12}(x, y, z)$ clearly reveals the non-diffracting property of the light field and its polarization singularities upon propagation. As both, the spatial frequency components of vertical as well as horizontal polarization, individually form a non-diffracting light field, their incoherent superposition naturally inherits this non-diffracting property but revealing the complex polarization structure including vectorial singularities in real space, being likewise non-diffracting: In 3d space, V-points observed in the light field's transverse planes form straight line singularities parallel to the optical axis. As highlighted in Fig. 3.12b, following the sign rule, neighboring singularities on a joint zero-line in 2d space are alternating in sign or a self-intersection of zero-lines is found in between. Note that, as underlined by the number of crossing zero-lines, found V-points are non-generic, thus, under minor perturbations, they unfold into pairs of generic C-points. However, due to the non-diffracting characteristic of the light field, only changing insignificantly due to minor perturbations, the C-point pairs would not diverge but propagate parallelly to the V-point trajectories in Fig. 3.12a. Hence, instead of V-points adopting the non-diffracting property, the corresponding C-points would inherit this propagation characteristic.

3.3.3 Vectorial Self-imaging Light Fields

Probably the best known phenomenon, in which the reproduction of transverse planes of a light field is observed, represents the Talbot effect. In 1836 H. F. Talbot firstly

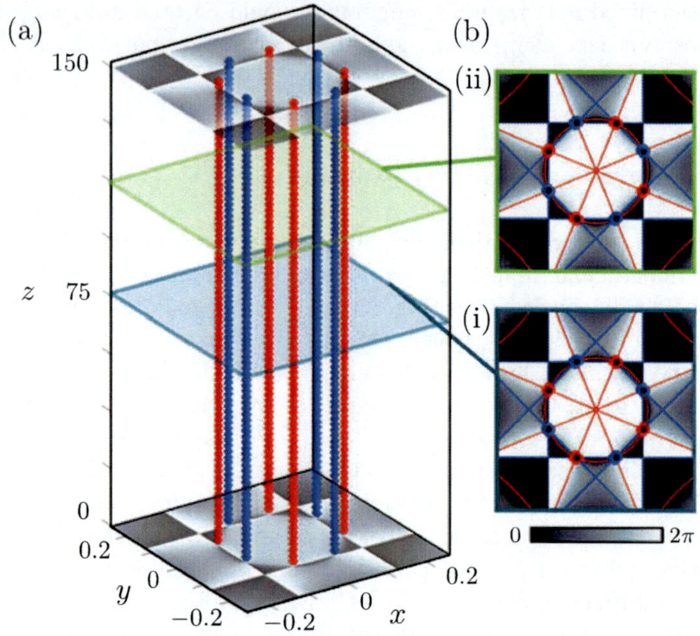

Fig. 3.12 Calculated propagation dynamics of polarization singularities in the vectorial discrete non-diffracting field of Fig. 3.11a–c. **a** Shown is a 3d plot of the singular points with $\sigma_{12} = \pm 2$ (red/blue dot: positively/negatively indexed) in $\Phi_{12}(x, y, z)$ found closest to the optical axis (x, y, z in mm). **b** Singularities (red/blue circle: positively/negatively indexed) are detected by zero-line intersections in Φ_{12} ($Z_{\Re,\Im}$: red, blue lines in (i) and (ii))

described this effect witnessed behind periodic diffractive objects of transverse lattice constant g [81]. Many years later in 1881 [82], L. Rayleigh analytically explained the observation and defined the Talbot length $d_{\text{Talbot}} = 2g^2/\lambda$ as the longitudinal period of the repeating transverse (scalar) field structure. More than a century after its first observation, in 1967 W. D. Montgomery generalized this effect of periodically reproduction of the complex transverse amplitude and phase, introducing "self-imaging" light fields [83].

Following the description in Ref. [52], the scalar self-imaging light field replicating with period d (in z-direction) is assumed by the ansatz

$$\begin{aligned} E_{\text{si}}(x, y, z) &= E_{\text{si}}(x, y, z + d) \\ &= \sum_m E_{\text{si},\perp}(x, y)\, e^{i\frac{2\pi m z}{d}}, \qquad m \in \mathbb{N}_0 \end{aligned} \tag{3.16}$$

with the transverse field $E_\perp(x, y)$. Combined with the scalar wave equation (Eq. (2.9)), this ansatz results in the characteristic 2d differential equation

$$\left[\frac{\partial^2}{\partial x^2}+\frac{\partial^2}{\partial y^2}+k^2\left\{1-\left(\frac{m\lambda}{d}\right)^2\right\}\right]\cdot E_{\mathrm{si},\perp}(x,y)=0. \tag{3.17}$$

Fourier transforming this equation reveals a specific characteristic of self-imaging light fields: In Fourier space, the spatial frequency components, which fulfill the self-imaging criteria, are found on circular rings of radius

$$\rho_m=\sqrt{\frac{1}{\lambda^2}-\left(\frac{m}{d}\right)^2} \tag{3.18}$$

where $m \in \mathbb{N}_0$ represents the number of the so-called Montgomery rings with $0 \leq m \leq m_{\max}$, $m_{\max} \leq d/\lambda$ for $\rho_m \in \mathbb{R}$. An exemplary self-imaging field including (a) its far-field distribution[12] as well as its repeating (b) intensity $s_0 \in [0, 1]$ and (c) phase $\phi \in [0, 2\pi]$ structure in 3d real space is illustrated in Fig. 3.13 for $z \in [0, d]$ (simulation with numerical approach following Ref. [78]; 9 slices). Based on the chosen symmetry and number of frequency components in k-space, representing six plane waves interfering with different $\vec{k}$-vectors in real space, a structure of hexagonal symmetry is formed at $z = 0$: Centrally, an arrangement of six intensity spots surrounds an on-axis intensity maxima in s_0. This configuration is again surrounded by six arrangements each constituted of three intensity maxima (and some relative maxima). The same distribution is found for $z = d$ where the light field is self-imaged. Upon propagation the intensity structure as well as phase clearly changes, whereby also in intermediate planes the general hexagonal symmetry is kept, as visible in, e.g., honeycomb-like intensity structures.

Generally, for self-imaging fields the Montgomery rings, i.e., their number as well as $m_{\max}$, and number of frequency components as well as their angular position and phase relation can be chosen independently, facilitating a great variety of scalar self-imaging light fields [78]. Interestingly, considering this property of self-imaging fields, non-diffracting light fields can be seen as a subclass of these, exploiting only a single Montgomery ring in Fourier space [78]. Till now, self-imaging fields have only been realized including a homogeneous polarization distribution. Similar to vectorial discrete non-diffracting light fields, here, the far-field distribution of a self-imaging field is proposed to be utilized for the innovative formation of its vectorial, polarization structured, analogon: Besides the DoFs available for scalar self-imaging light fields (see above), the polarization is added as an additional DoF resulting in polarization structured light fields in real space. Hence, an incoherent or partially coherent superposition of tailored self-imaging light fields will be observed,[13] namely

[12] Also in this case, ideally, the frequency components represent points in Fourier/k-space (the far-field). For experimental purposes, these are approximated by circles as indicated in Figs. 3.13a and 3.14a

[13] Note that this assumption is only correct in the paraxial regime, i.e., if a lens of low numerical aperture (NA) is applied for the optical Fourier transformation from far- (Fourier/k-space) to near-field (real space).

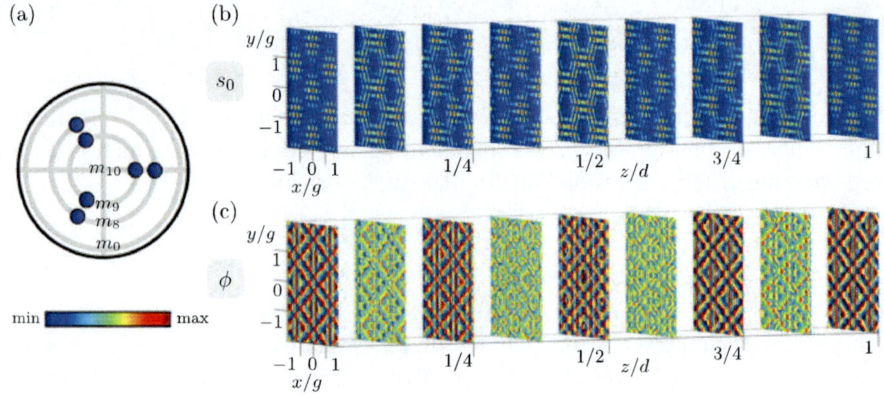

Fig. 3.13 Exemplary scalar self-imaging light field. **a** Six spatial frequency components are chosen without phase difference and positioned on two Montgomery rings ($m_{8,9}$ with $m_j = j$, $j = \{0, 1, 2, ..., m_{\max}\}$, $m_{\max} = 10$) in Fourier space. **b**, **c** Respective real space distribution of the self-imaging field indicated by nine z-slices of the intensity $s_0 \in [0, 1]$ and phase $\phi \in [0, 2\pi]$ ($z \in [0, d]$). (Numerical calculation following Ref. [78])

$$\vec{E}_{\mathrm{si}}(x, y, z) = \sum_{n=1}^{N} \vec{e}_n \cdot E_{\mathrm{si},n}(x, y, z) \quad (3.19)$$

if N different states of polarization $\vec{e}_n$ were chosen in k-space. Therefore, a vectorial light field sculpted in all three spatial dimensions is formed.

3.3.3.1 The 3d Structured Vectorial Field

As representative example for vectorial self-imaging fields a configuration of 12 spatial frequency components is chosen without phase difference and positioned on two Montgomery rings ($m_{8,9}$ with $m_j = j$, $j = \{0, 1, 2, ..., m_{\max}\}$, $m_{\max} = 10$) in Fourier space as shown in Fig. 3.14a. Even though in general various states of polarization could be chosen for the individual frequency components, as an example two orthogonal states are chosen, namely vertically and horizontally linear polarization ($\vec{e}_1 = [1, 0]^T, \vec{e}_2 = [0, 1]^T, N = 2$ in Eq. (3.19)). In the chosen configuration, neighboring frequency components on a joint ring are of orthogonal polarization, whereas nearest neighbors on different rings have the same polarization (see arrows). Thus, the resulting setting incoherently superimposes two self-imaging light fields, only differing in angular orientation and polarization. Experimentally, this superposition is realized by far-field construction (experimental system: Fig. 3.10) and analyzed by a spatially resolved Stokes parameter measurement system as introduced in Sect. 3.3.2.

In real space, this superposition results in a complex structured field, as visualized in Fig. 3.14b by calculated (top) and measured (bottom) Stokes parameters $s_0 \in [0, 1]$ and $S_{1,2,3} \in [-1, 1]$. Obviously, experimental results nicely confirm the

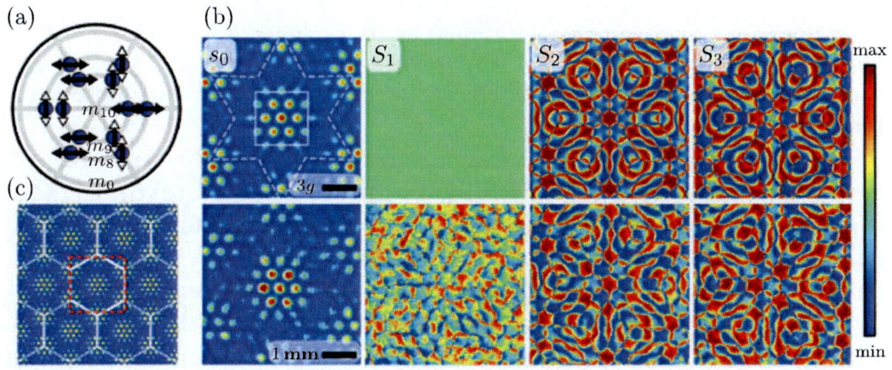

Fig. 3.14 Formation of exemplary vectorial self-imaging light field. **a** Configuration in Fourier space and **b** resulting theoretical (top) and experimentally measured (bottom) intensity $s_0 \in [0, 1]$ and polarization structure (Stokes parameters $S_{1,2,3} \in [-1, 1]$) in real space for $z = 0$. The distributions are of repeating hexagonal symmetry as highlighted by the six-fold star (white dashed line) in **b** and white lines in the zoom-out of s_0 in **c** (red dashed square in **c**: area studied in **b**; white square in **a**: area studied in Fig. 3.16)

theoretical calculations. Again, deviations between experiment and theory are due to unfolding vectorial singularities (details in following section) as well as inaccuracies and polarization sensitivity of optical components. Clearly, the intensity configuration s_0 resembles the intensity structure of the scalar field presented in Fig. 3.13: Six intensity spots are found surrounding an on-axis intensity maxima. Hexagonally arranged around this and connected by a star-like pattern (see dashed white line in Fig. 3.14b, s_0) six intensity configurations are observed, each consisting of three intensity spots surrounded by some sub-maxima. The similarity between the intensity of the scalar self-imaging field and this vectorial version at $z = 0$ is due to the fact that the latter represents an incoherent superposition of the presented scalar field plus this scalar field rotated by 60° (see Fig. 3.14a). As these two scalar fields show the same transverse intensity distribution and their relative phase difference is translated into spatially varying polarization states, s_0 of the vectorial self-imaging field has the same appearance as its basis scalar fields.

Approximately, the shown area of the real space light field presents a unit cell of the overall transverse structure. This is indicated in Fig. 3.14c depicting the honeycomb arrangement of these hexagonal cells in a broader s_0 area of the light field. Beyond, considering the polarization structure in the transverse plane, the hexagonal symmetry is confirmed, clearly visible in S_2 and S_3 in Fig. 3.14b. Hence, the respective symmetry in all substructures in real space is strongly related to the symmetry in k-space.

Note that, even though only linear states of polarization are applied in the far-field, in real space the respective vectorial field represents an ellipse field being sculpted in S_2 and S_3, whereas $S_1 = 0$. This behavior is easily explained by the applied superposition principle: Both, the horizontally as well as vertically linearly polarized scalar self-imaging field, being combined in real space, include a phase configuration, as exemplified in Fig. 3.13c. The according relative spatial phase difference between both orthogonally polarized fields is translated into states of polarization of transversely and longitudinally varying ellipticity, hence, a complex 3d ellipse field is formed.

The study of this vectorial self-imaging field in 3d space is shown in Fig. 3.15. Here, the propagation of normalized Stokes parameters (a) S_1, (b) S_2, and (c) S_3 is presented by nine transverse z-slices (simulation) within the distance $z \in [0, d]$, i.e., within a self-imaging distance of both underlying scalar fields of orthogonal polarization. Below each 3d plot, experimental results are presented for chosen z-distances. The white box in Fig. 3.15a indicates the area investigated in the 3d plots.

For the experimental investigation of the light field $\vec{E}_{\mathrm{si}}(x, y, z)$ in 3d (real) space *digital propagation* [84] is additionally applied. Equivalent to the appoach used for the numerical simulation of light field propagation (cf. Eq. (3.5)), this experimental method is based the angular spectrum [85, 86], according to which a propagated scalar field is given by

$$E(x, y, z) = \mathcal{F}^{-1}\left\{\mathcal{F}\left[E(x, y, 0)\right] \cdot \mathrm{e}^{\mathrm{i}k_z z}\right\}. \tag{3.20}$$

Hence, by the application of Fourier hologram for the light field of interest (for $z = 0$) in combination with phase shifts $\pm k_z z$ ($k_z^2(k_x, k_y) = k^2 - k_x^2 - k_y^2$), encoded on a phase-only SLM, the respective light field can be digitally propagated in $\pm z$-direction. In the case at hand, this digital propagation method is performed for both incoherently superimposed self-imaging fields ($E_{\mathrm{si},1,2}$), so that the propagated field $\vec{E}_{\mathrm{si}}(x, y, z)$ for a chosen z is created in the real space observation plane. For this purpose the respective phase shifts $k_z(k_x, k_y)z$ are implemented on the first half of the SLM of the experimental system (cf. Fig. 3.10a, b1). By this approach, the realized vectorial self-imaging field can be investigated for different propagation distances z without moving the detection/measurement system.

As proven by the 3d plots and perfectly confirmed by the experimental results, the polarization configuration of the vectorial field shows a similar self-imaging behavior as its scalar analogon. As characteristic for self-imaging, the same polarization structure is observed for $z = 0$ and $z = d$ (cf. Fig. 3.15a–c). This property is based by the fact that both superimposed scalar self-imaging fields $E_{\mathrm{si},1,2}$ (cf. Eq. (3.19)) possess the same self-imaging distance d. For the named self-imaged and all intermediate planes the described hexagonal symmetry is kept with a smooth transition between these planes. For the planes at $z = \{0, d\}$ as well as $z = d/2$ the polarization structure is purely modulated in S_2 and S_3, whereas in between these planes all three Stokes parameters reveal an inhomogeneous distribution. This behavior is due to z-dependent changes in relative phase between the two superimposed self-imaging fields of orthogonal polarization. In total, a complex polarization structure is realized, which spatially varies in all three spatial dimensions and inherits self-imaging characteristics of its underlying scalar bases.

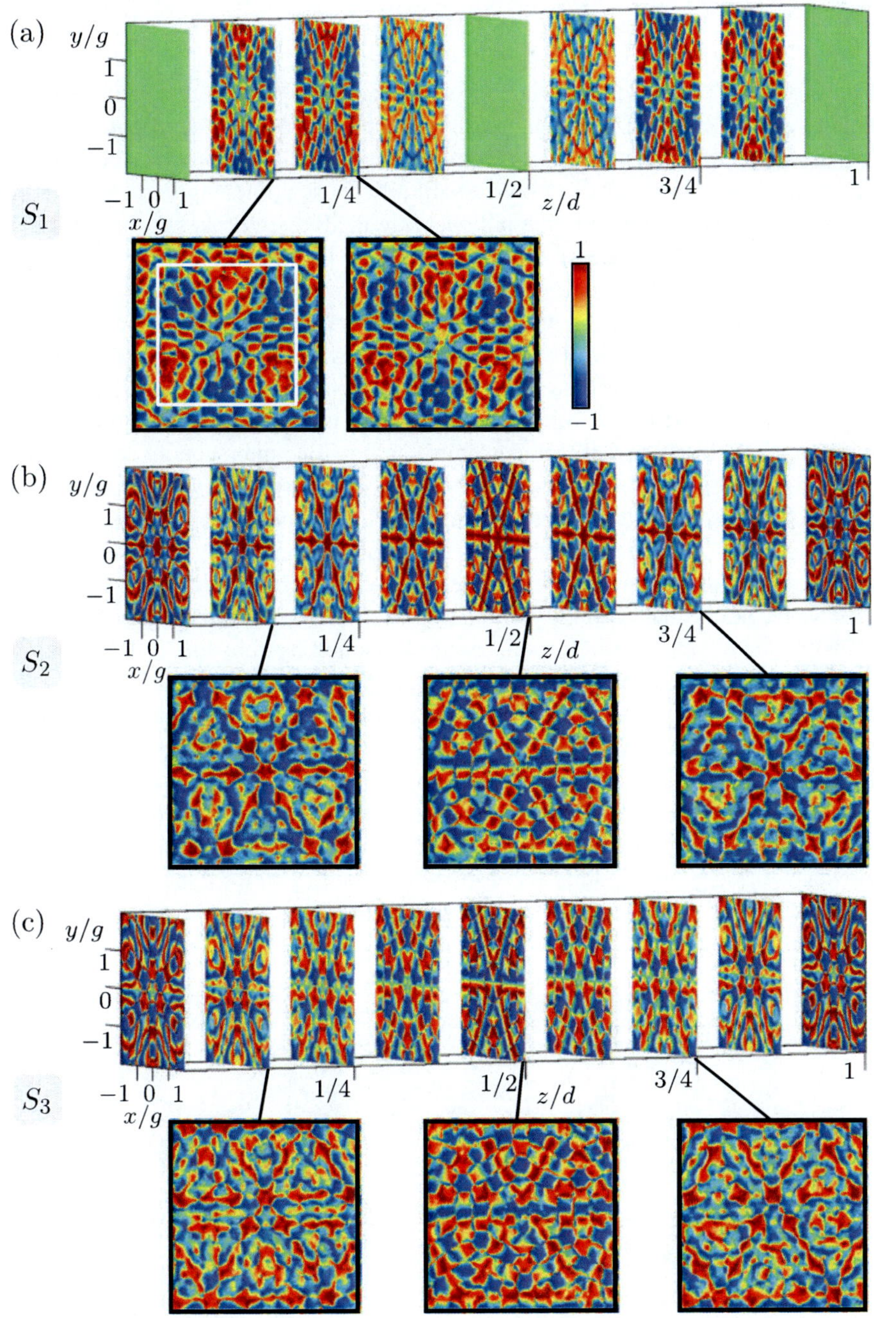

Fig. 3.15 Self-imaging polarization configuration in 3d space. Shown are z-slices of the normalized Stokes parameters **a** S_1, **b** S_2, and **c** S_3 of the exemplary vectorial self-imaging light field (Fig. 3.14) for $z \in [0,\ d]$. Top: numerical results in 3d plot, bottom: experimental results with white box in **a** indicating the area investigated numerically

3.3.3.2 Singularity Dynamics in 3d Space

Considering the complex polarization structure of the vectorial self-imaging light field in 3d space, polarization singularities are expected in this configuration. Here, the 3d ellipse field is examined for vectorial point singularities in its z-slices, namely non-generic V- and generic C-points. Naturally, these point singularities form lines in 3d space. Since the field is spatially varying in all three Stokes parameters $S_{1,2,3}$, singularities are identified by the Stokes field phases Φ_{23} (V-points) and Φ_{12} (C-points).

Figure. 3.16 shows the analysis of the 0-plane ($z = 0$) of the exemplary vectorial self-imaging field (chosen area marked by white box in Fig. 3.14b). For the study of V-point singularities, the phase Φ_{23} is examined (a) theoretically and (b) experimentally, calculated from measured Stokes parameters. Both, simulation and measurement clearly reveal singular points as zero-line intersections with $Z_{\Re}$ (red) and $Z_{\Im}$ (blue) corresponding to $S_2 = 0$ and $S_3 = 0$, respectively. Detected singular points of positive/negative singularity index σ_{23} are marked by red/blue circles. As visible in Fig. 3.16a, simulations show a SN of non-generic V-points, each found as a crossing point of multiple zero-lines. Thereby, the arrangement of singularities follows the hexagonal symmetry of the transverse field. Nearest neighbors on a joint zero-line are of opposite index sign or are separated by zero-line self-intersections (saddle points). Thus, the sign rule is kept. Confirming the non-genericity of respective V-points, experimental results reveal unfolded singularities at the position of numerically determined singularities. This observance is due to the V-point inherent instability combined with experimental constraints of, e.g., applying circles instead of points within the far-field construction. As a consequence, stationary points of linear polarization with $\sigma_{23} = \pm 1$ are found instead of V-points with $\sigma_{23} = \pm 2$. However, these generic points also form a SN adapting to the expected symmetry.

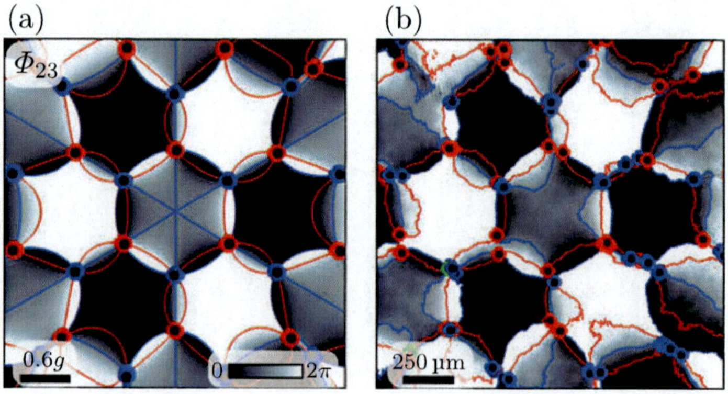

Fig. 3.16 V-point polarization singularity analysis by Stokes field phase Φ_{23} in the $z = 0$ plane of the self-imaging field—**a** simulation, **b** experiment. Zero-lines $Z_{\Re}/Z_{\Im}$ are highlighted red/blue with positively/negatively indexed Φ_{23}-singularities marked by red/blue circles

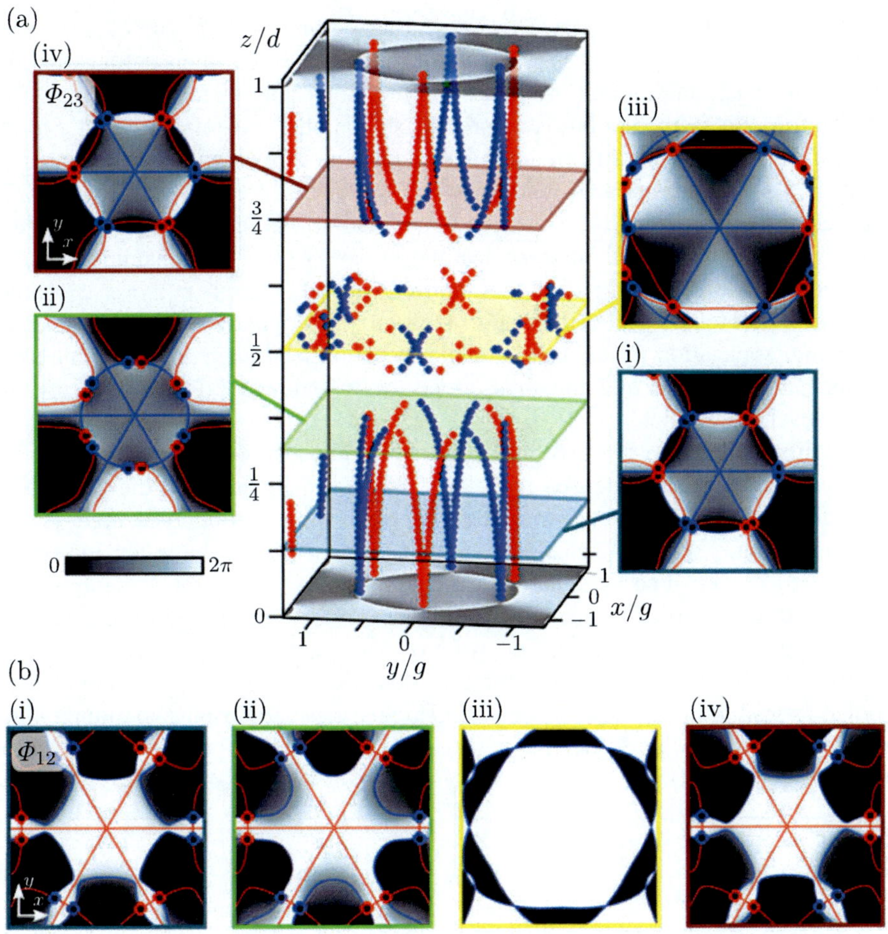

Fig. 3.17 Polarization singularity dynamics in exemplary vectorial self-imaging field analyzed in 3d space. **a** 3d plot of positively/negatively indexed singularities (red/blue dots) identified in $\Phi_{23}(x, y, z)$. In (i)–(iv) insets of exemplary transverse planes are shown with red/blue $Z_{\Re}/Z_{\Im}$ zero-lines and singularities marked by red/blue circles (positive/negative index). **b** Insight into C-point creation, convergence, and annihilation in 3d space studied by phase $\Phi_{12}(x, y, z)$ with respective zero-lines (red, blue: $Z_{\Re,\Im}$) and marked by red/blue circles (positive/negative index)

Upon propagation, these point singularities form lines (in 3d space) exhibiting the creation and annihilation of singular points (in 2d space) following the complex self-imaging structure. The respective numerical analysis is presented in Fig. 3.17a, showing a 3d plot of detected Φ_{23}-singularities (red/blue dots: positively/negatively indexed). Here, the inner hexagon of singularities in Fig. 3.16a, i.e., in the 0-plane, was chosen for detailed studies. Insets (i)–(iv) present exemplary transverse planes of Φ_{23} with zero-lines $Z_{\Im,\Re}$ (blue, red) and marked singular points (red/blue circles: positively/negatively indexed). Going from bottom to top, i.e., starting at the self-imaged 0-plane, six transverse non-generic V-points with $\sigma_{23} = \pm 2$ are found, unfolding into generic stationary points of index $\sigma_{23} = \pm 1$ upon propagation (cf. (i)) following (total) singularity index conservation. Note that this splitting is not due to minor perturbation but is tailored by shaping the self-imaging field. Shortly after the plane shown in (ii), singularities of opposite index annihilate each other, so that no Φ_{23}-singularities are found. Around $z = d/2$, the creation (before $z = d/2$), subsequent convergence (see (iii)), and annihilation (behind $z = d/2$) of singular points is observed. Reflecting the repeating characteristic of self-imaging fields, the creation of stationary points and their convergence to V-points at $z = d$ is evinced for $z > d/2$. Hence, while adhering to the sign and singularity index conservation rule, the propagation dynamics of singularities is mirror-symmetric with the plane of mirror being at $z = d/2$ and, thus, adopts the customized self-imaging properties of the overall light field.

Additionally, Fig. 3.17b gives an insight into embedded Φ_{12}-singularities, i.e., C-points in 2d space, and their propagation dynamics in 3d space. At the position (i), the afore occurred creation of transverse C-points in pairs of opposite sign, located on a joint (blue) $Z_{\mathfrak{I}}$ zero-line ($S_2 = 0$), is evinced. Note the C-points considered on a joint (red) $Z_{\mathfrak{R}}$ zero-line are of the same index, so that a saddle point (self-intersection of $Z_{\mathfrak{R}}$) is found in between. All C-points are of first order ($\sigma_{12} = \pm 1$; generic) as concluded from the standard crossing of one $Z_{\mathfrak{R}}$ and one $Z_{\mathfrak{I}}$ zero-line. Upon propagation, singularities of opposite sign slowly converge till they annihilate each other, as visible in (iii) with no detected singularities. Confirming the longitudinal mirror-symmetry of the light field, in (iv) the re-creation of C-points is found similar to planes (i) and (ii).

3.3.4 Conclusion

Within this section an innovative approach is proposed for the realization of complex vectorial light fields customized in 3d space and inheriting valuable properties as non-diffraction, self-healing, or self-imaging. Besides demonstrating the formation of non-diffracting vector Bessel-Gaussian beam by a combination of SLM and q-plate, the property of self-healing within non-diffracting vector fields was evinced by an all-digital approach [15]. Significantly enlarging the variety of known vectorial non-diffracting fields, the novel diverse class of vectorial discrete non-diffracting light fields was introduced and experimentally realized by far-field construction. Besides being non-diffracting in amplitude, phase and polarization, these fields embed polarization SNs also inheriting the non-diffracting property of the underlying scalar ancestors. In the frame of their study, a sophisticated metrological technique was developed enabling the phase analysis of light fields being of inhomogenous polarization. Beyond, vectorial self-imaging light fields were proposed, proving the not yet observed combination of transverse as well as longitudinal on-demand customization of fully-structured fields by far-field construction. The transverse shape of the field is strongly dependent on chosen parameters and symmetry in k-space. Crucially, adapting to this symmetry, a network of polarization singularities is shaped. Upon propagation, the singularity unfolding, annihilation, and subsequent (re)creation with repeating character in z-direction is evinced. These till now inaccessible polarization singularity dynamics reveal the pathbreaking adaptability of singularity propagation properties by the presented approach, even allowing for self-imaging singularity characteristics.

References

1. Berry MV, Dennis MR (2001) Knotted and linked phase singularities in monochromatic waves. Proc R Soc Lond A 457:2251
2. Berry MV, Dennis MR (2001) Knotting and unknotting of phase singularities: Helmholtz waves, paraxial waves and waves in 2+1 spacetime. J Phys A: Math Gen 34:8877
3. Dennis MR et al (2010) Isolated optical vortex knots. Nat Phys 6:118
4. Bliokh KY et al (2008) Singular polarimetry: Evolution of polarization singularities in electromagnetic waves propagating in a weakly anisotropic medium. Opt Express 16:695
5. Fang Y et al (2017) Fractional-topological-charge-induced vortex birth and splitting of light fields on the submicron scale. Phys Rev A 95:023821
6. Leach J et al (2005) Vortex knots in light. New J Phys 7:55
7. Dennis MR (2002) Polarization singularities in paraxial vector fields: morphology and statistics. Opt Commun 213:201
8. Cardano F et al (2013) Generation and dynamics of optical beams with polarization singularities. Opt Express 21:8815
9. Lopez-Mago D et al (2013) Dynamics of polarization singularities in composite optical vortices. J Opt 15:044028
10. Vyas S, Kozawa Y, Sato S (2013) Polarization singularities in superposition of vector beams. Opt Express 21:8972
11. Lu X et al (2014) Polarization singularities and orbital angular momentum sidebands from rotational symmetry broken by the Pockels effect. Sci Rep 4:4865
12. Ram BB, Sharma A, Senthilkumaran P (2017) Diffraction of V-point singularities through triangular apertures. Opt Express 25:0270
13. Flossmann F et al (2005) Polarization singularities from unfolding an optical vortex through a birefringent crystal. Phys Rev Lett 95:253901
14. Otte E, Alpmann C, Denz C (2018) Polarization singularity explosions in tailored light fields. Laser Photon Rev 12:1700200
15. Otte E et al (2018) Recovery of nonseparability in self-healing vector Bessel beams. Phys Rev A 98:053818
16. Otte E, Denz C (2018) Sculpting complex polarization singularity networks. Opt Lett 43:5821
17. Dennis MR, O'Holleran K, Padgett MJ (2009) Singular optics: optical vortices and polarization singularities. In: Wolf, E. (ed.) Progress in optics, vol. 53. Chap. 5, pp. 293–363. Elsevier
18. Freund I (2002) Polarization singularity indices in Gaussian laser beams. Opt Commun 201:251
19. Yu R et al (2013) Array of polarization singularities in interference of three waves. J Opt Soc Am A 30:2556
20. Pal SK, Ruchi S, Senthilkumaran P (2017) C-point and V-point singularity lattice formation and index sign conversion methods. Opt Commun 393:156
21. Pal SK, Senthilkumaran P (2018) Lattice of C points at intensity nulls. Opt Lett 43:1259
22. Ruchi S, Pal K, Senthilkumaran P (2017) Generation of V-point polarization singularity lattices. Opt Express 25:1932
23. Kurzynowski P, Woźniak WA, Borwińska M (2010) Regular lattices of polarization singularities: their generation and properties. J Opt 12:035406

24. Kurzynowski P et al (2012) Singularities of interference of three waves with different polarization states. Opt Express 20:26755
25. Liang C et al (2017) Vector optical coherence lattices generating controllable far-field beam profiles. Opt Express 25:9872
26. Milione G et al (2015) 4x 20 Gbit/s mode division multiplexing over free space using vector modes and a q-plate mode (de)multiplexer. Opt Lett 40:1980
27. Ndagano B et al (2017) Characterizing quantum channels with non-separable states of classical light. Nat Phys 13:397
28. Bandres MA, Gutiérrez-Vega J (2004) Ince Gaussian beams. Opt Lett 29:144
29. Alpmann C et al (2017) Dynamic modulation of Poincaré beams. Sci Rep 7:8076
30. Terhalle B (2011) Controlling light in optically induced photonic lattices. Springer Science & Business Media
31. Malik M et al (2012) Influence of atmospheric turbulence on optical communications using orbital angular momentum for encoding. Opt Express 20:13195
32. Rodenburg B et al (2012) Influence of atmospheric turbulence on states of light carrying orbital angular momentum. Opt Lett 37:3735
33. Andrews LC, Phillips RL (2005) Laser beam propagation through random media, vol 152. SPIE press Bellingham
34. Leader JC (1978) Atmospheric propagation of partially coherent radiation. JOSA 68:175
35. Cox MA et al (2016) On the resilience of scalar and vector vortex modes in turbulence. Opt Express 24:18105
36. Soskin MS, Vasnetsov MV (2001) Singular optics. In: Wolf E (ed) Progress in optics, vol 42, Chap 4. Elsevier, pp 219–277
37. Rubinsztein-Dunlop H et al (2017) Roadmap on structured light. J Opt 19:013001
38. Karimi E et al (2007) Hypergeometric-Gaussian modes. Opt Lett 32:3053
39. Fahrbach FO, Simon P, Rohrbach A (2010) Microscopy with self-reconstructing beams. Nat Photon 4:780
40. Planchon TA et al (2011) Rapid three-dimensional isotropic imaging of living cells using Bessel beam plane illumination. Nat Methods 8:417
41. Fahrbach FO, Rohrbach A (2012) Propagation stability of self-reconstructing Bessel beams enables contrast-enhanced imaging in thick media. Nat Commun 3:632
42. Dubey AK, Yadava V (2008) Laser beam machining - a review. Int J Mach Tools Manuf 48:609
43. McGloin D, Dholakia K (2005) Bessel beams: diffraction in a new light. Contemp Phys 46:15
44. McGloin D, Garcés-Chávez V, Dholakia K (2003) Interfering Bessel beams for optical micromanipulation. Opt Lett 28:657
45. Arlt J et al (2008) Optical micromanipulation using a Bessel light beam. Opt Commun 197:239
46. Volke-Sepulveda K et al (2002) Orbital angular momentum of a high-order Bessel light beam. J Opt B: Q Semiclass Opt 4:S82
47. Broky J et al (2008) Self-healing properties of optical Airy beams. Opt Express 16:12880
48. Ring JD et al (2012) Auto-focusing and self-healing of Pearcey beams. Opt Express 20:18955
49. Bouchal Z (2002) Resistance of nondiffracting vortex beam against amplitude and phase perturbations. Opt Commun 210:155
50. Mendoza-Hernández J et al (2015) Laguerre-Gauss beams versus Bessel beams showdown: peer comparison. Opt Lett 40:3739
51. Aiello A et al (2017) Unraveling beam self-healing. Opt Express 25:19147
52. Patorski K (1989) The self-imaging phenomenon and its applications. In: Wolf E (ed) Progress in optics, vol 27. Elsevier, pp 1–108
53. Bouchal Z, Kyvalský J (2004) Controllable 3D spatial localization of light fields synthesized by non-diffracting modes. J Mod Opt 51:157
54. Rogers G (1969) Fourier images in electron microscopy and their possible misinterpretation. J Microsc 89:121
55. Ulrich R, Kamiya T (1978) Resolution of self-images in planar optical waveguides. JOSA 68:583
56. Blume S (1976) Analogies between electronic and optical systems. Optik 46:333

57. Garcia-Gracia H, Gutiérrez-Vega JC (2016) Polarization singularities in nondiffracting Mathieu-Poincaré beams. J Opt 18:014006
58. He M et al (2013) Propagation properties and self-reconstruction of azimuthally polarized non-diffracting beams. Opt Commun 294:36
59. Dudley A et al (2013) Generating and measuring nondiffracting vector Bessel beams. Opt Lett 38:3429
60. Wu G, Wang F, Cai Y (2014) Generation and self-healing of a radially polarized Bessel-Gauss beam. Phys Rev A 89:043807
61. Milione G et al (2015) Measuring the self-healing of the spatially inhomogeneous states of polarization of vector Bessel beams. J Opt 17:035617
62. Davis JA et al (2016) Nondiffracting vector beams where the charge and the polarization state vary with propagation distance. Opt Lett 41:227
63. Moreno I et al (2015) Nondiffracting Bessel beams with polarization state that varies with propagation distance. Opt Lett 40:5451
64. Li P et al (2017) Generation and self-healing of vector Bessel-Gauss beams with variant state of polarizations upon propagation. Opt Express 25:5821
65. Durnin J, Miceli JJ, Eberly JH (1499) Diffraction-free beams. Phys Rev Lett 58: 1499
66. Durnin J (1987) Exact solutions for nondiffracting beams. I. The scalar theory. JOSA A 4:651
67. Gori F, Guattari G, Padovani C (1987) Bessel-Gauss beams. Opt Commun 64:491
68. Litvin IA, McLaren MG, Forbes A (2009) A conical wave approach to calculating Bessel-Gauss beam reconstruction after complex obstacles. Opt Commun 282:1078
69. McGloin D, Dholakia K (2005) Bessel beams: diffraction in a new light. Contemp Phys 46:15
70. Bouchal Z, Wagner J, Chlup M (1998) Self-reconstruction of a distorted nondiffracting beam. Opt Commun 151:207
71. Turunen J, Vasara A, Friberg AT (1988) Holographic generation of diffraction-free beams. Appl Opt 27:3959
72. Cottrell DM, Craven JM, Davis JA (2007) Nondiffracting random intensity patterns. Opt Lett 32:298
73. Marrucci L, Manzo C, Paparo D (2006) Optical spin-to-orbital angular momentum conversion in inhomogeneous anisotropic media. Phys Rev Lett 96:163905
74. Piccirillo B et al (2010) Photon spin-to-orbital angular momentum conversion via an electrically tunable q-plate. Appl Phys Lett 97:241104
75. Nape I et al (2018) Self-healing high-dimensional quantum key distribution using hybrid spin-orbit Bessel states. Opt Express 26:26946
76. Boguslawski M, Rose P, Denz C (2011) Increasing the structural variety of discrete nondiffracting wave fields. Phys Rev A 84:013832
77. Greco V, Molesini G, Quercioli F (1995) Accurate polarization interferometer. Rev Sci Instrum 66:3729
78. Alpmann C (2015) Tailored light fields - Holographic modulation of amplitude, phase and polarization. PhD thesis. University of Muenster
79. Kreis T (1986) Digital holographic interference-phase measurement using the Fouriertransform method. JOSA A 3:847
80. Schaefer B et al (2007) Measuring the Stokes polarization parameters. Am J Phys 75:163
81. Talbot HF (1836) LXXVI. Facts relating to optical science. No. IV. Lond Edinb Dubl Philos Mag J Sci 9: 401
82. Rayleigh L (1881) XXV. On copying diffraction-gratings, and on some phenomena connected therewith. Lond Edinb Dubl Philos Mag J Sci 11: 196
83. Montgomery WD (1967) Self-imaging objects of infinite aperture. JOSA 57:772
84. Otte E et al (2018) Entanglement beating in free space through spin-orbit coupling. Light Sci Appl 7: 18009 (cit. on pp. g, 27, 28, 78, 118, 130-142)
85. Goodman J (2008) Introduction to Fourier optics. McGraw-Hil
86. Rosales-Guzmán C, Forbes A (2017) How to shape light with spatial light modulators. SPIE Press

Chapter 4
Non-paraxial 3d Polarization in 4d Light Fields

Recently, research has mainly been focused on structured singular light in the paraxial regime, as also presented in previous chapters. Here, investigations were concentrated on fields shaped in 2d or 3d space, in which the typically elliptical polarization of light is purely transverse, i.e., two-dimensional (2d polarization).[1] Hence, the light field oscillates solely within the plane orthogonal to the beam's propagation direction with $\vec{E}(x, y, z) = [E_x(x, y, z), E_y(x, y, z)]^T$. Respective studies have lead to pathbreaking findings enabling advanced applications as well as new insights into the general and not yet fully explored, nor understood, nature of light (see Chaps. 2, 3). However, a major current challenge is the moving from these 3d fields of 2d polarization to 4d structured light of 3d polarization with $\vec{\mathcal{E}}(x, y, z) = [\mathcal{E}_x(x, y, z), \mathcal{E}_y(x, y, z), \mathcal{E}_z(x, y, z)]^T$, exploiting all three electric field components, i.e., transverse as well as longitudinal ones. These fields do not only give new insights into the fundamentals of light, but also open new horizons for applied optics: 4d fields allow for structures of nano-scale complexity whereby their 3d polarization nature is an enriching feature for handling, e.g., polarization sensitive materials. The crucial longitudinal components are non-negligible for non-paraxial light fields, fulfilling the vectorial Helmholtz equation (Eq. (2.7)), as in tightly focused beams. These fields are formed by the application of focusing optical components imparting a numerical aperture (NA) larger or equal to 0.7. In this case, radial components of the input light field are transformed into significant longitudinal focal field contributions [1–3]. Considering this effect, some studies have been performed representing first steps towards the investigation and future application of 4d structured fields. For example, tightly focusing 2d polarization structures enabled shaping focal intensity configurations [4–8] as optical needles. Furthermore, longitudinal field components facilitate the formation of complex 3d polarization topologies

[1] Here, even though longitudinal contributions might exist, their contribution is neglecable compared to transverse ones.

E. Otte, *Structured Singular Light Fields*, Springer Theses,
https://doi.org/10.1007/978-3-030-63715-6_4

surrounding optical singularities as optical cones, twisted ribbons and Möbius strips. These topologies were as firstly predicted by Freund in 2005 [9] and, ten years later, experimentally proven by Bauer et al. [10].

This chapter concentrates on the demanded advancement of the studies on 4d fields, including its customization, application, experimental identification, and contribution to singular optics. First, the fundamentals for the study of non-paraxial fields are presented (Sect. 4.1). Following, a method for the customization of non-paraxial intensity landscapes in 2d and 3d space is proposed (Sect. 4.2), revealing the nano-scale complexity of non-paraxial 4d light. The benefit of this method is demonstrated by optically trapping zeolite-L nano-containers, giving insights into future applications in the field of polarization sensitive nano-assemblies. Although 4d structured fields may pave the way to disruptive applications, their experimental realization and, particularly, customization is still hampered as their fast experimental analysis is impeded by the 3d polarization nature and nano-scale complexity. Solving this issue by combining nano-chemistry and -optics, an innovative nano-tomographic technique is proposed allowing for the single-shot identification of non-paraxial 4d fields (Sect. 4.3). Finally, studies are focused on optical singularities as well as 3d polarization topologies in tailored non-paraxial fields (Sect. 4.4), even introducing not yet observed Möbius strip arrays.

Note that the results presented in this chapter have in parts been published in Refs. [4, 11–13].

4.1 From Paraxial to Non-paraxial Light Fields

For customizing 4d structured light fields in all its degrees of freedom, namely, amplitude, phase as well as 3d polarization, strongly focusing sculpted paraxial fields represents a mighty tool. The 4d nature as well as the nano-scale complexity achievable within these non-paraxial light fields include pathbreaking potential for the advancement of applications as in nano-plasmonics, optical manipulation, or the implementation of 4d functional nano-materials (cf. Sect. 4.3). Here, precisely shaped light fields are required which are adaptable to the 4d, polarization sensitive, and/or nano-scale nature of the applications. For the developement of appropriate fields, the in-depth theoretical understanding of strongly focused light is demanded, leading to its optimal implementation. Here, the analytical approach [14] and, based on this, the further applied numerical technique [15] is introduced enabling the thorough investigation of tailored 4d light fields.

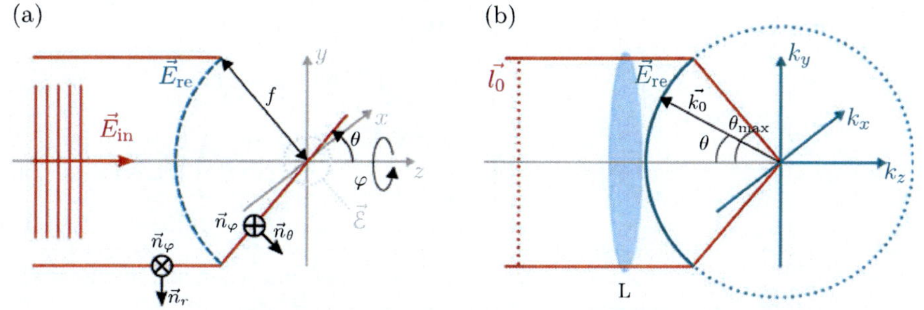

Fig. 4.1 Calculation of focal field distributions. Conceptional sketch for **a** analytical and **b** numerical approach. By means of a refracting reference sphere (blue dashed line), the incident paraxial field $\vec{E}_{in}$ or $\vec{l}_0$ is transformed into the focal field (gray) $\vec{\mathcal{E}}$ (in gray real space coordinates (x, y, z) or (φ, θ, z)) via the refracted field (blue) $\vec{E}_{re}$ (in blue Fourier/k-space coordinates (k_x, k_y, k_z)). A high-NA lens (L) of focal distance f is applied. ($(r, \varphi)/(\varphi, \theta)$: polar/spherical coordinates; $\vec{n}_{r,\varphi}/\vec{n}_{\varphi,\theta}$: unit vectors of the cylindrical/spherical coordinate system)

4.1.1 Analytical Calculation of Focal Fields

To determine the focusing properties of light, one can apply Richards and Wolf's integrals which are based on electromagnetic diffraction in optical systems [14]. The following respective description is based on the derivation in Ref. [16]. A paraxial input light field is assumed to be focused by an aplanatic optical lens of focal distance f, as illustrated in Fig. 4.1a. Considering a reference sphere of radius f refracting the incident light rays to the focal point, the non-paraxial field can be calculated. By means of the sphere the incident light field $\vec{E}_{in}(r, \varphi) = [E_x, E_y, E_z]^T(r, \varphi)$ (red) with $E_z \approx 0$ and its transverse cylindrical coordinates (r, φ) are transformed into the refracted field $\vec{E}_{re}(\theta, \varphi)$ (blue) with spherical coordinates (θ, φ), which subsequently leads to the focal field $\vec{\mathcal{E}}(r, \varphi, z)$ (gray). For this purpose, the input field is split into its s- and p-polarized parts, namely

$$\vec{E}_{in}^{s} = \left(\vec{E}_{in} \cdot \vec{n}_\varphi\right) \cdot \vec{n}_\varphi, \qquad \vec{E}_{in}^{p} = \left(\vec{E}_{in} \cdot \vec{n}_r\right) \cdot \vec{n}_r \tag{4.1}$$

with the unit vectors $\vec{n}_{r,\varphi}$ of the cylindrical coordinate system. While unit vector $\vec{n}_\varphi$ is unaffected by refraction, $\vec{n}_r$ is mapped into the spherical unit vector $\vec{n}_\theta$ (cf. Fig. 4.1a). Hence, the total refracted electric field behind the optical lens can be conveniently calculated from

$$\vec{E}_{re}(\theta, \varphi) = \left[t^s\left(\vec{E}_{in} \cdot \vec{n}_\varphi\right)\vec{n}_\varphi + t^p\left(\vec{E}_{in} \cdot \vec{n}_r\right)\vec{n}_\theta\right]\sqrt{\cos\theta} \tag{4.2}$$

with Fresnel transmission coefficients $t^{s,p}$ [17]. By expressing unit vectors in terms of Cartesian unit vectors $\vec{n}_{x,y,z}$ with spherical coordinates, the equation above converts into

$$\vec{E}_{\text{re}}(\theta,\varphi) = t^s(\theta)\left[\vec{E}_{\text{in}}(\theta,\varphi)\cdot\begin{pmatrix}-\sin\varphi\\ \cos\varphi\\ 0\end{pmatrix}\right]\begin{pmatrix}-\sin\varphi\\ \cos\varphi\\ 0\end{pmatrix}\sqrt{\cos\theta}$$
$$+\, t^p(\theta)\left[\vec{E}_{\text{in}}(\theta,\varphi)\cdot\begin{pmatrix}\cos\varphi\\ \sin\varphi\\ 0\end{pmatrix}\right]\begin{pmatrix}\cos\varphi\cos\theta\\ \sin\varphi\cos\theta\\ -\sin\theta\end{pmatrix}\sqrt{\cos\theta}. \quad (4.3)$$

Applying angular spectrum representation[2] with substitutions of spatial frequencies $k_{x,y,z}$ and transverse (real space) coordinates $(x,\ y)$ according to

$$k_x = k\ \sin\theta\ \cos\varphi, \qquad k_y = k\ \sin\theta\ \sin\varphi, \qquad k_z = k\ \cos\theta,$$
$$x = r\ \cos\varphi, \qquad y = r\ \sin\varphi$$

results in the final expression for the focal field in polar coordinates with

$$\vec{\mathcal{E}}(r,\ \varphi,\ z) = -\frac{ikf\,\mathrm{e}^{-ikf}}{2\pi}\int_0^{\theta_{\max}}\int_0^{2\pi}\vec{E}_{\text{re}}(\theta,\varphi)\,\mathrm{e}^{ikz\cos\theta}\,\mathrm{e}^{ikr\sin\theta\cos(\theta-\varphi)}\sin\theta\,\mathrm{d}\varphi\,\mathrm{d}\theta. \quad (4.4)$$

Here, the dependence of the focusing lens on the numerical aperture NA is found within the maximal angle (semi aperture angle) $0 < \theta_{\max} < \pi/2$ with

$$\mathrm{NA} = n_{\text{re}}\sin\theta_{\max} \quad (4.5)$$

and the refractive index n_{re} of the surrounding medium. The derived Eq. (4.4) in combination with Eq. (4.3) now enables the analytical calculation of arbitrary non-paraxial fields created by tightly focusing structured input light fields. For instance, focal field distributions were calculated resulting from focusing a homogeneously polarized field [16, 18], radial and azimuthal cylindrical vector beams (CVBs) [2, 19, 20], or (on-axis) superpositions of these [3, 5–7, 21]. Solving respective vectorial diffraction integrals analytically is still feasibly as long as input light fields represent, e.g., exact analytical solutions of the wave equation. Here, respective transverse symmetry allows for the application of Bessel functions for solving the integration over the variable φ (e.g., see Ref. [16]). Thus, the overall integration is reduced to the variable θ with a collection of typical integrals recurring for various different input fields.

Consequently, there are clear constraints of the analytical approach as for input fields of spatially tailored amplitude, phase as well as polarization the computational effort will increase dramatically. Therefore, numerical techniques for solving the diffraction integrals represent a valuable tool since they facilitate the focal field calculation for arbitrarily structured incident light fields. In the following, the numerical approach applied within this thesis is introduced.

---

[2]Note that within this step the refracted field $\vec{E}_{\text{re}}$ is assumed to be the far field distribution of the focal electric field.

4.1.2 Numerical Approach for Focusing Light

For the calculation of non-paraxial 4d light fields formed by focusing a complex structured field, one can solve vectorial diffraction integrals numerically by means of fast Fourier transfrom (FFT) operations.[3] The following short description of this approach is based on Ref. [15], in which the technique was derived and demonstrated in detail. For this method, the customized inhomogeneously polarized input field $\vec{E}_{\text{in}}$, i.e., pupil function $\vec{l}_0 = [l_{0,x}, l_{0,y}]^T$, is separated into two orthogonally polarized basis $l_{0,x}$ and $l_{0,y}$. As a consequence, the refracted electric field at the exit aperture of the focusing optics (Fig. 4.1b) is given by

$$\vec{E}_{\text{re}}(\theta, \varphi) = \sqrt{\cos\theta}\left[l_{0,x} \begin{pmatrix} \cos\theta\cos^2\varphi + \sin^2\varphi \\ (\cos\theta - 1)\cos\varphi\sin\varphi \\ -\sin\theta\cos\varphi \end{pmatrix} + l_{0,y} \begin{pmatrix} (\cos\theta - 1)\cos\varphi\sin\varphi \\ \cos\theta\sin^2\varphi + \cos^2\varphi \\ -\sin\theta\sin\varphi \end{pmatrix} \right]. \tag{4.6}$$

As described below, based on this representation and Eq. (4.4), focus calculations are performed for the x- (1st summand) and y-polarized (2nd summand) part independently. Subsequently, the overall non-paraxial light field is calculated from the sum of both results.

To obtain focal field results in real space coordinates while applying 2d Fourier transform ($\mathcal{F}$) for its determination, the refracted light field is considered in k-space, as sketched in Fig. 4.1b (blue). For this purpose, spherical coordinates θ and φ are substituted according to

$$\cos\theta = k_z/k_0, \quad \sin\theta = k_r/k_0, \quad \cos\varphi = k_x/k_r, \quad \sin\varphi = k_y/k_r,$$
$$k_z^2 = k_0^2 - k_r^2, \quad k_r^2 = k_x^2 + k_y^2, \quad k_0 \sin\theta_{\max} = 1.$$

Considering these and Eq. (4.4), the total non-paraxial field $\vec{\mathcal{E}}(x, y, z)$ in Cartesian coordinates (x, y, z) can be derived as

$$\begin{aligned} \vec{\mathcal{E}}(x, y, z) &= \mathcal{F}\left[\vec{E}_{\text{re}}(k_x, k_y) \cdot \mathrm{e}^{\mathrm{i}k_z z} \cdot k_z/k_0\right] \\ &= \mathcal{F}\left[l_{0,x}(k_x, k_y) \cdot \vec{GX}(k_x, k_y)\right] + \mathcal{F}\left[l_{0,y}(k_x, k_y) \cdot \vec{GY}(k_x, k_y)\right], \end{aligned} \tag{4.7}$$

with the 2d Fourier transform $\mathcal{F}$ and the $\vec{G}$-functions

$$\vec{GX}(k_x, k_y) = \mathrm{e}^{\mathrm{i}k_z z}\sqrt{\frac{k_0}{k_z}} \begin{pmatrix} \frac{k_0 k_y^2 + k_z k_x^2}{k_0 k_r} \\ \frac{(k_z - k_0)k_x k_y}{k_0 k_r} \\ -\frac{k_x}{k_0 k_r} \end{pmatrix}, \tag{4.8}$$

[3] Within this thesis numerical calculations were performed utilizing the software MATLAB by MathWorks.

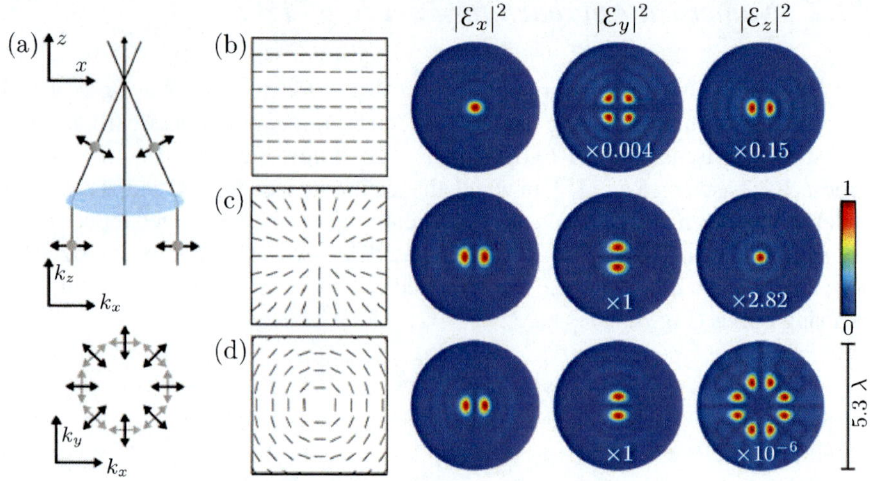

Fig. 4.2 Exemplary focal fields formed by tightly focusing optics. **a** Concept of the creation of longitudinal focal field components (bottom: incident field, top: focusing concept; black/gray arrow or dots: radial/azimuthal components). **b**–**d** Focusing properties (NA = 0.9) of a x-, radially and azimuthally polarized input field (left). Resulting focal intensity contributions $|\mathcal{E}_{x,y,z}|^2$ are shown on the right, each normalized to its own maximum with its ratio to the maximum of $|\mathcal{E}_x|^2$ shown within images

$$\vec{GY}(k_x, k_y) = \mathrm{e}^{\mathrm{i}k_z z}\sqrt{\frac{k_0}{k_z}}\begin{pmatrix}\frac{k_0 k_y^2 + k_z k_x^2}{k_0 k_r} \\ \frac{(k_z - k_0)k_x k_y}{k_0 k_r} \\ \frac{k_x}{k_0 k_r}\end{pmatrix} \tag{4.9}$$

(see Ref. [15] for details). Here, $\vec{GX}$ and $\vec{GY}$ include the focusing properties of the high-NA lens acting on the input light field $\vec{l}_0$. Fourier transformations within Eq. (4.7) can be conveniently calculated by the numerical FFT operation. To enhance resolution of the calculated focal field distributions, the incident light field is embedded in a larger matrix of zeros (zero padding).

Figure 4.2 shows the concept of creating 3d polarization and well-known examples of focal fields determined numerically by the FFT based approach. Here, Fig. 4.2a illustrates the creation of longitudinal field components by tightly focusing (top) radial field components (black arrows) of the incident light field (bottom). In contrast, azimuthally oriented components (gray arrows/dots) are unaffected when passing a high-NA lens. Note that coordinate systems in Fig. 4.2a indicate the transformation of the incident field $\vec{l}_0 = \vec{E}_{\text{in}}$ via the refracted field in k-space to the focal field $\vec{\mathcal{E}} = [\mathcal{E}_x, \mathcal{E}_y, \mathcal{E}_z]^T$ in real space with coordinates (x, y, z). As first example, the focusing properties of a homogeneously linearly polarized field is presented in Fig. 4.2b. In Fig. 4.2c and d the two extreme cases of pure radial or azimuthal polarization structure as input light field are demonstrated. All input fields are without any amplitude or phase shaping, thus, pure polarization structures are chosen. In each subfigure (b)–(d), on the left, the incident light field is illustrated with polarization being represented by black lines. On the right, for a numerical aperture of NA $= 0.9$ (in air, refractive index $n_{\mathrm{re}} = 1$) the transverse ($|\mathcal{E}_{x,y}|^2$) and longitudinal ($|\mathcal{E}_z|^2$) focal field contributions to the total intensity $|\vec{\mathcal{E}}(x, y, 0)|^2$ are illustrated for $z = 0$, i.e., within the focal plane. Each intensity distribution is normalized to its own maximum, whereby the relation between x-, y- and z-contributions is given within the images. Here, the given factor represent the ratio of the maximum of $|\mathcal{E}_{y,z}|^2$ to the maximum of $|\mathcal{E}_x|^2$.

For homogeneous input polarization oriented horizontally, i.e., $\vec{E}_{\mathrm{in}}(x, y) = [1, 0]^T$, the focal field is mainly constituted of x-components forming a central spot. However, also minor y- and z-contributions are found due to the mixture of radial and azimuthal components within the transverse plane of the incident light field. Since the incident field includes pure radial components on the center's left and right, two intensity spots are formed by longitudinal focal field components $\mathcal{E}_z$. In contrast, if the complete transverse plane of the incident field is polarized radially (Fig. 4.2c), strong E_z components are created. Here longitudinal components form a Gaussian spot of significant contribution whose maximum is almost three-times ($\times 2.82$) as high as the ones of transverse contributions. If a field of pure azimuthal polarization is focused (Fig. 4.2d), the focal field can be assumed as purely transverse with neglectable longitudinal field contributions.

Note that the presented numerical approach is applied throughout this chapter for calculating and, thus, studying focal 4d fields.

4.2 Tailored Non-paraxial Intensity Structures

Above the two extreme cases of focusing a purely radial or azimuthal vector beam were presented. Input fields including both, radial and azimuthal components can be realized easily by superposition of basic radial and azimuthal fields. However, in this case, the amount of radial and azimuthal components is constant within the whole transverse plane resulting in a spatially homogeneous distribution of focal longitudinal field contributions. In contrast, based on the knowledge of radial components being responsible for the creation of longitudinal contributions and azimuthal ones supporting transverse focal field components, here a sophisticated customization approach is proposed for sculpting focal landscapes of spatially inhomogeneous 3d polarization. For this purpose, the ratio of radial and azimuthal electric field components of the input light field will be shaped within the transverse plane such that, as a consequence, the spatial structure of longitudinal and transverse focal field contributions is tailored [4]. Beyond, the approach will be extended by additional phase modulation of the incident light field enabling a broad variety of tailored 4d fields [12]. To prove and study this customization tool in detail, different 4d light fields will be calculated in the following dependent on the structured input field. Besides investigating the focal plane, the complete 3d volume of customized 4d fields are explored. Beyond, the potential of created landscapes for future applications is proven by optical manipulation of cylindrical nano-containers [11].

4.2.1 The Focal Plane

4.2.1.1 Bright Flowers and Dark Stars

First, the focal 2d plane of the tailored 4d fields $\vec{\mathcal{E}}(x, y, 0)$ is investigated. For spatially shaping the ratio of incident radial and azimuthal components, higher-order vector fields represent a perfect means [4]. As presented in Sect. 2.3.2, depending on the index σ_{12} of the embedded V-point singularity, the transverse polarization distribution is structured. As indicated in Fig. 4.3a/c with $\sigma_{12} = \pm 8$, for positive/negative indices σ_{12} a vectorial flower/web structure with $|\sigma_{12} - 2|$ petals/sectors is formed. Crucially, the number of petals/sectors is connected to the number $N = |\sigma_{12} - 2|$ of radially oriented ζ-lines being of pure radial polarization. Hence, by adjustment of σ_{12} the amount of purely radially oriented states of polarization and, thus, the spatial structure of the total focal field and its transverse and longitudinal field contributions can be sculpted. This effect is demonstrated in Fig. 4.3b or d, showing the focal field distribution for focusing the flower or web structure in Fig. 4.3a or b, respectively, with an numerical aperture of NA = 0.9 (in air, refractive index $n_{\text{re}} = 1$). Here, the total focal intensity $|\vec{\mathcal{E}}|^2 \in [0, 1]$, the intensity contributions $|\mathcal{E}_{x,y,z}|^2 \in [0, 1]$ and the respective relative phase distributions $\phi_{x,y,z} \in [0, 2\pi]$ are depicted, revealing the 3d polarization nature of the 4d non-paraxial light field. Each intensity structure is

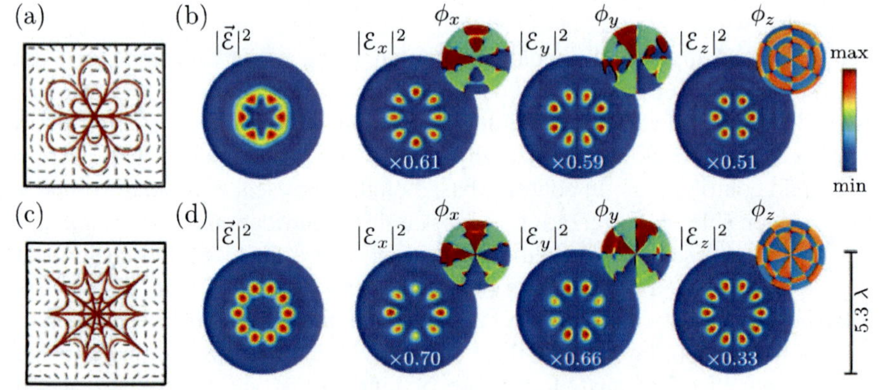

Fig. 4.3 Shaping the focal field by higher-order vector fields. Incident vectorial **a** flower and **c** web structure (red: flow lines) form a **b** dark star and **d** bright flower as total focal intensity $|\vec{\mathcal{E}}|^2 \in [0, 1]$ (NA = 0.9). Respective intensity contributions $|\mathcal{E}_{x,y,z}|^2 \in [0, 1]$ are shown with corresponding phase structures $\phi_{x,y,z} \in [0, 2\pi]$. [Adapted with permission from Ref. [4], licensed under CC BY 4.0, https://creativecommons.org/licenses/by/4.0/]

normalized to its own maximum whereby the ratio (peak ratio) of the maximum of $|\mathcal{E}_{x,y,z}|^2$ to the maximum of $|\vec{\mathcal{E}}|^2$ is given within each image.

The total focal intensity forms a $|\sigma_{12} - 2| = 6$-point dark star in its center (b) for the focused flower ($\sigma_{12} = 8$), whereas for the focused web structure ($\sigma_{12} = -8$) a $|\sigma_{12} - 2| = 10$-fold bright flower configuration is sculpted (d). The created shapes are due to the focal field contributions $\mathcal{E}_{x,y,z}$ being structured according to the ratio of incident radial to azimuthal components and the respective ζ-lines: Following the number N and position of ζ-lines, the intensity structure of the longitudinal contribution reveals $N = |\sigma_{12} - 2|$ intensity spots surrounding the optical axis. Thus, a petal configuration is formed with (b) 6 or (d) 10 petals in $|\mathcal{E}_z|^2$. Accordingly, the phase structures ϕ_z reveal discrete phase values of $\pi/2$ and $3\pi/2$, with a phase difference of π between neighboring intensity petals. Similarly, the transverse contributions $|\mathcal{E}^2_{x,y}|$ form petal configurations following the spatially varying amount of radial and azimuthal input components. Here, $|\sigma_{12}|$ intensity spots are observed again going along with a discrete phase structure showing phase differences of π between neighboring petals. Note that due to differences in Gouy phase shifts for x-/y- and z-components [22], a relative phase difference of $\pi/2$ is identified between $\phi_{x,y}$ and ϕ_z.

The peak ratio of $\mathcal{E}_{x,y,z}$ reveal a strong contribution of longitudinal field components with a ratio of $\times 0.51$ for the focused flower (b) and $\times 0.33$ for the focused web structure (d). Consider that the peak ratio for Fig. 4.3d is lower than for Fig. 4.3b since $|\mathcal{E}_z|^2$ includes more petals for the focused web. Furthermore, for both, the focused flower and web, the transverse contribution $\mathcal{E}_{x,y}$ include a maximum intensity which is approximately 0.6 to 0.7 times as high as the maximum of the total intensity. Note that, although the peak ratios of z-contributions seem to be equal (b) or even small (d) compared to transverse ones, the spatial mean of the longitudinal intensity distribution reveal to be about half as strong as the ones of transverse parts, as it will be shown in the following. Summed up, the transverse and longitudinal contributions with its different peak ratios form the respective dark star or bright flower (Table 4.1).

Results above indicate the ability to customize the focal 4d field by adapting the singularity index σ_{12} of the incident singular vector field or, more precisely, its (2d) polarization singularity [4]. In Table 4.1 the focal intensity landscapes including its transverse and longitudinal contributions are presented for varying singularity indices $|\sigma_{12}| = \{4, 6, 8, 10, 12\}$ of incident paraxial (a) flower or (b) web configurations. Also in this case, intensity distributions are normalized to its own maxima ($|\vec{\mathcal{E}}|^2$, $|\mathcal{E}_{x,y,z}|^2 \in [0, 1]$). Thereby the peak ratio ("peak") as well as mean ratio ("mean"), i.e., the ratio of the mean value of $|\mathcal{E}_{x,y,z}|^2$ to the mean of $|\vec{\mathcal{E}}|^2$, are shown below each image. Studying the different intensity distributions confirms the findings above, i.e., for a focused flower or web structure a $|\sigma_{12} - 2|$-point dark star ($\sigma_{12} > 4$) or $|\sigma_{12} - 2|$-fold bright flower is formed for $|\vec{\mathcal{E}}|^2$, respectively. Beyond, transverse contributions consist of $|\sigma_{12}|$ petals. Hence, by adjusting the singularity index σ_{12} of the incident field the focal field structure can be customized. Note that with varying index the peak and mean ratio also changes.

Table 4.1 Shaping focal intensity landscapes of 4d fields by tightly focusing vectorial (a) flowers or (b) spider webs with varying index σ_{12} of embedded singularity. The total focal intensity distribution $|\vec{\mathcal{E}}|^2$ resembles $|\sigma_{12} - 2|$-point stars/-fold flowers. Peak and mean ratio of each intensity contributions $|\mathcal{E}_{x,y,z}|^2$ is shown below respective images. [Adapted with permission from Ref. [4], licensed under CC BY 4.0, https://creativecommons.org/licenses/by/4.0/]

(a) Input flower to focal dark star

σ_{12}	$\lvert\vec{\mathcal{E}}\rvert^2$	$\lvert\mathcal{E}_x\rvert^2$	$\lvert\mathcal{E}_y\rvert^2$	$\lvert\mathcal{E}_z\rvert^2$
4				
peak		×0.76	×0.61	×0.63
mean		×0.400	×0.378	×0.222
6	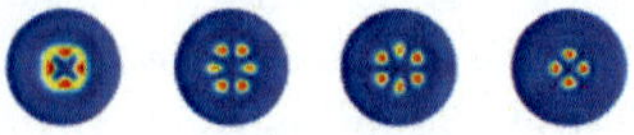			
peak		×0.62	×0.62	×0.55
mean		×0.389	×0.389	×0.222
8	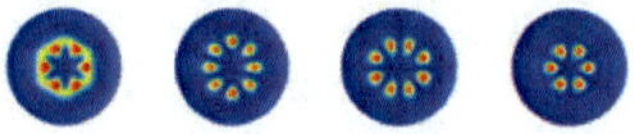			
peak		×0.61	×0.59	×0.51
mean		×0.386	×0.387	×0.226
10	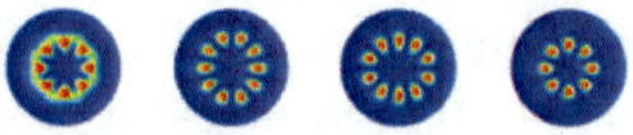			
peak		×0.56	×0.56	×0.50
mean		×0.385	×0.385	×0.230
12	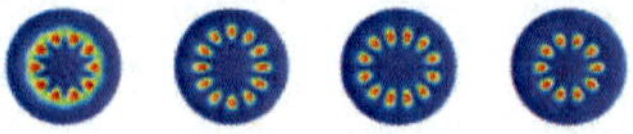			
peak		×0.56	×0.54	×0.49
mean		×0.383	×0.384	×0.233

(b) Input spider web to focal bright flower

σ_{12}	$\lvert\vec{\mathcal{E}}\rvert^2$	$\lvert\mathcal{E}_x\rvert^2$	$\lvert\mathcal{E}_y\rvert^2$	$\lvert\mathcal{E}_z\rvert^2$
−4	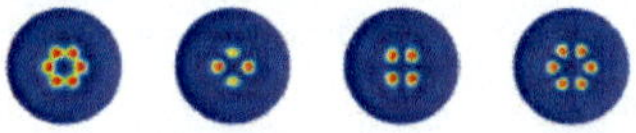			
peak		×0.75	×0.67	×0.30
mean		×0.388	×0.390	×0.222
−6	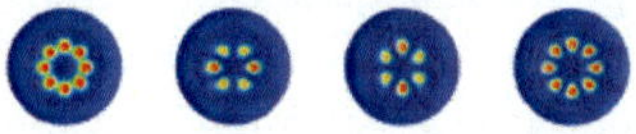			
peak		×0.72	×0.72	×0.32
mean		×0.388	×0.388	×0.223
−8	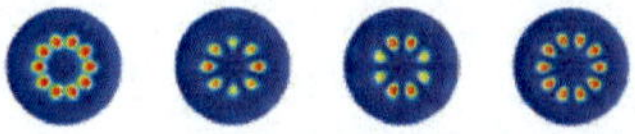			
peak		×0.70	×0.66	×0.33
mean		×0.387	×0.388	×0.225
−10	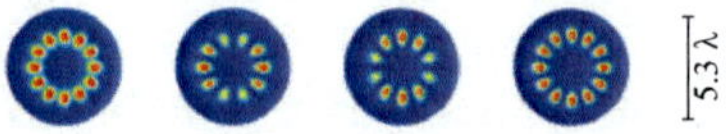			
peak		×0.68	×0.68	×0.34
mean		×0.386	×0.386	×0.229
−12	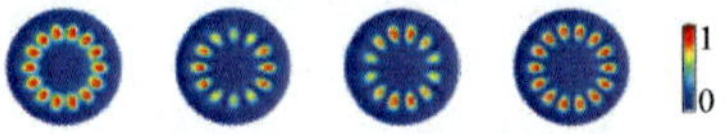			
peak		×0.67	×0.65	×0.34
mean		×0.384	×0.384	×0.232

For an increasing singularity index $|\sigma_{12}|$ the diameter of the intensity structures $|\vec{\mathcal{E}}|^2$ or $|\mathcal{E}_{x,y,z}|^2$ increase as well, whereby the central dark area expands simultaneously. Generally, the longitudinal components represent approximately 22% to 23% and transverse ones about 38% to 40% of the total focal intensity, as shown by mean ratio values. As a consequence of the increasing number $|\sigma_{12} - 2|$ of radially polarized ζ-lines of the input field, the mean ratio of $|\mathcal{E}_z|^2$ slightly increases with the index $|\sigma_{12}|$. Vice versa, the mean ratio of transverse contribution decreases. Thus, by adjusting σ_{12} the amount of longitudinal and transverse contributions as well as the overall spatial structure of the focal intensity landscapes can be tailored on demand.

4.2.1.2 Phase Modulation as Customization Tool

Above, pure polarization modulation has already proven its significant potential for shaping focal 4d fields of tailored 3d polarization. As demonstrated by the creation of singularity explosions in Sect. 3.2, additional global phase structuring may have a significant effect on the total tailored light field. Thus, here, polarization combined with phase modulation is applied on the input light field for customizing the focal 4d field [12]. For this purpose, the input field is chosen similarly to the paraxial singular light field of Sect. 3.2.1, Eq. (3.6), embedding the sophisticated tailored singularity splitting, i.e.,

$$\vec{E}_{\text{in}} = \left[\cos\left(\frac{\sigma_{12}}{2}\cdot\varphi\right),\ \sin\left(\frac{\sigma_{12}}{2}\cdot\varphi\right)\right]^T \cdot e^{i\ell\varphi}. \tag{4.10}$$

Crucially, these paraxial fields combine a vectorial with a scalar singularity, which does not only have a significant impact on the focal intensity landscapes but also singularities embedded in the according 4d singular field, as it will be demonstrated in Sect. 4.4.

As an illustrative example, similar to Fig. 4.3, $\sigma_{12} = \pm 8$ is chosen. Onto the according vectorial flower and web structure additional phase modulation is applied in the form of a phase vortex of charge $\ell = \pm\{0, 1, 2, ..., 5\}$. Table 4.2 presents the resulting non-paraxial 4d fields at $z = 0$ when the polarization and phase structured input field is focused ((a) $\sigma_{12} = 8$, (b) $\sigma_{12} = -8$) with an numerical aperture of NA $= 0.9$. Here, the total focal intensity $|\vec{\mathcal{E}}|^2$ as well as the intensity contributions $|\mathcal{E}_{x,y,z}|^2$ are shown, each normalized to its own maximum. Again, the peak and mean ratio is included below each image.

Results show that properties of the total focal intensity $|\vec{\mathcal{E}}|^2$ as well as intensity contributions $|\mathcal{E}_{x,y,z}|^2$ vary depending on chosen topological charge $|\ell|$. For instance, with increasing $|\ell|$, the transverse size of intensity configurations firstly decreases ($|\ell| \leq 4$), then increases again ($|\ell| > 4$). Furthermore, the values of the peak and mean ratios are changing. Interestingly, for $|\ell| = |\sigma_{12}|/2 = 4$ the creation of a Gaussian-like transverse structure for $|\vec{\mathcal{E}}|^2$ is observed in both cases, the focused flower and web configuration. Note that by simulations of focal fields for various σ_{12} this observance has been proven. For the Gaussian-like $|\vec{\mathcal{E}}|^2$ distribution the focal transverse contributions also form Gaussian-like spots of major impact on the total structure, whereas a relatively small peak ratio is found for $|\mathcal{E}_z|^2$. Further, the intensity landscapes for the tightly focused flower with additional vortex of charge $\ell = \pm 3$ (± 5) resemble the structures for the focused web with vortex of charge $\ell = \pm 5$ (± 3). In these cases, a major longitudinal contribution of Gaussian shape is formed for $|\ell| = |\sigma_{12} - 1|$. In contrast, for $|\ell| = |\sigma_{12} + 1|$ contributions of z-polarization are significantly smaller and of donut shape (Table 4.2).

Hence, by the application of additional phase variation on the input light field as customization tool, not only the shape of the resulting non-paraxial 4d field is sculpted, but also the ratio of transverse to longitudinal components can be tailored by the choice of ℓ in addition to σ_{12}. Beyond, the additional phase vortices have a

Table 4.2 Customization of non-paraxial 4d fields by additional global phase variation of the incident vectorial (a) flower and (b) web configuration ($\sigma_{12} = \pm 8$, NA = 0.9). Additional phase modulation represent vortices of charge $\ell = \pm\{0, 1, 2, ..., 5\}$. Shown are the total focal intensity $|\vec{\mathcal{E}}|^2$ as well as intensity contributions $|\mathcal{E}_{x,y,z}|^2$ of transverse and longitudinal polarization components for $z = 0$. Peak and mean ratio values are given below intensity images

(a) Focused vectorial flower structure

ℓ	$\lvert\vec{\mathcal{E}}\rvert^2$	$\lvert\mathcal{E}_x\rvert^2$	$\lvert\mathcal{E}_y\rvert^2$	$\lvert\mathcal{E}_z\rvert^2$
0	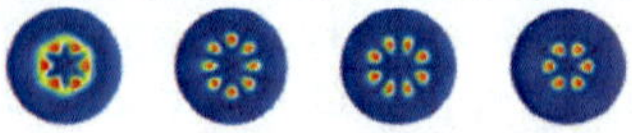			
peak		×0.60	×0.59	×0.52
mean		×0.383	×0.391	×0.226
±1	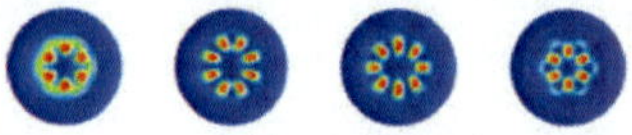			
peak		×0.49	×0.50	×0.39
mean		×0.383	×0.390	×0.227
±2	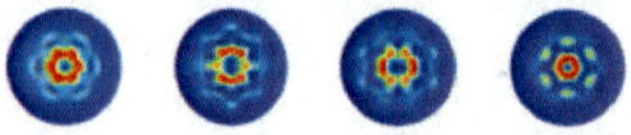			
peak		×0.46	×0.48	×0.39
mean		×0.384	×0.389	×0.227
±3	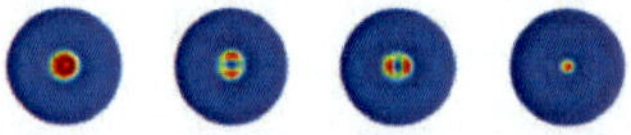			
peak		×0.57	×0.57	×0.83
mean		×0.382	×0.389	×0.228
±4				
peak		×0.49	×0.51	×0.07
mean		×0.380	×0.389	×0.231
±5				
peak		×0.48	×0.49	×0.16
mean		×0.379	×0.386	×0.235

(b) Focused vectorial web structure

ℓ	$\lvert\vec{\mathcal{E}}\rvert^2$	$\lvert\mathcal{E}_x\rvert^2$	$\lvert\mathcal{E}_y\rvert^2$	$\lvert\mathcal{E}_z\rvert^2$
0	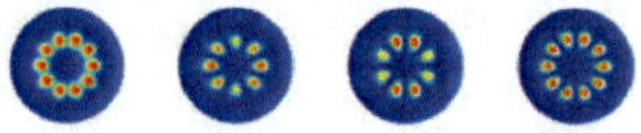			
peak		×0.69	×0.67	×0.33
mean		×0.384	×0.391	×0.225
±1	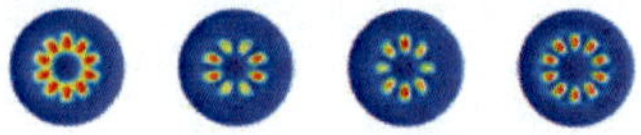			
peak		×0.66	×0.70	×0.32
mean		×0.384	×0.390	×0.226
±2	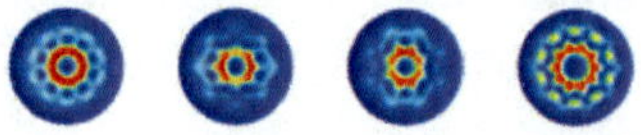			
peak		×0.50	×0.50	×0.22
mean		×0.384	×0.388	×0.227
±3	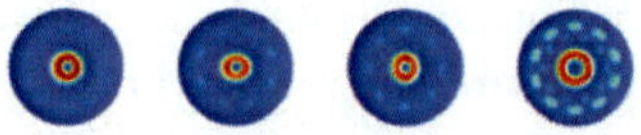			
peak		×0.49	×0.49	×0.16
mean		×0.382	×0.389	×0.228
±4	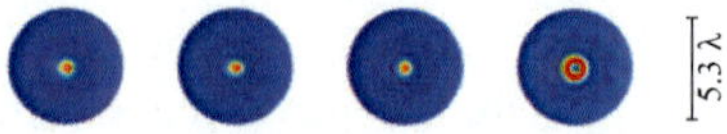			
peak		×0.49	×0.51	×0.07
mean		×0.381	×0.389	×0.229
±5	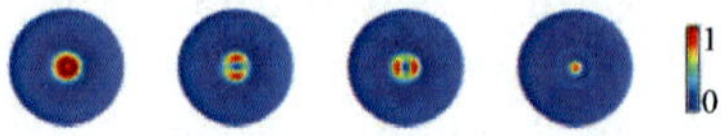			
peak		×0.56	×0.57	×0.83
mean		×0.381	×0.387	×0.231

major impact on the structure of the 4d field in 3d space, i.e., within the focal volume, which will be demonstrated in the following.

4.2.2 *Customizing the Focal 3d Volume*

Besides customizing the transverse focal plane of the complex 4d field, i.e., $\vec{\mathcal{E}}(x, y, 0)$ ($z = 0$), also the total focal volume of the non-paraxial light field, i.e., $\vec{\mathcal{E}}(x, y, z)$, is shaped by the modulation of the input field [11]. Considering applications, e.g., in optical micromanipulation, the customization of the complete focal volume is of special interest as it affects the properties of the optical trap (cf. Sect. 4.2.3).

To study the effect of structuring the input field on the focal volume, the 3d intensity landscapes $|\vec{\mathcal{E}}(x, y, z)|^2$ and its contribution $|\mathcal{E}_{x,y,z}(x, y, z)|^2$ are calculated for tightly focusing (NA = 0.9) a vectorial flower structure with $\sigma_{12} = 8$ and additional phase vortices of exemplary charge $\ell = \{0, 1, 3\}$. Numerically determined results are presented in Fig. 4.4. Here, results in the first row correspond to the focused flower without phase variation, the second and third row represent results for the focused flower with additional vortex of charge $\ell = 1$ and $\ell = 3$, respectively. In each row, z-slices of the (a)/(c)/(e) total 3d intensity and of (b)/(d)/(f) its contributions are shown in a 3d plot for $z \in [-7z_R, 7z_R]$ ($z_R \approx 0.25\lambda$, z_R: Rayleigh range). All intensity distributions are normalized to its own maximum with the peak ratio listed below 3d plots in Fig. 4.4b, d, and f. The blue and red surfaces within the plots mark values of (a), (b) 0.1/(c)–(f) 0.2 and (a)–(f) 0.8, respectively.

For the focused flower configuration, the total dark star intensity (a) as well as its contributions (b) keep its transverse shape upon propagation, i.e., within the different z-planes. Comparable to self-similar Gaussian modes, the transverse distributions enlarge their diameter and lose intensity with distance from the focal plane with $z = 0$ (bluish complete plane in 3d plots). In z-direction, a strong intensity gradient is observed as for $|z| = 2z_R$ the intensity already lost about 20% of its original maximum. Note that strong intensity gradients are desirable for optical trapping applications, since a major part of the trapping force, namly the gradient force [23–25], is proportional to the spatial intensity gradient. The (a) total intensity has a tube-like shape in 3d space, whereby the structure and diameter of the tube are adjustable by the choice of index σ_{12}. If, in contrast, a negative index would be chosen, a transverse bright flower would build the basis of the tube. Naturally, by adapting the absolute value $|\sigma_{12}|$, the number of dark star points or bright flower petals can be tailored, resulting in the adaption of the tube diameter (cf. Table 4.1).

If a phase vortex and, thus, orbital angular momentum (OAM) of charge $\ell = 1$ is added to the input light field, slight changes compared to the pure focused flower are observed (Fig. 4.4c–f). Still, a structure similar to a transverse dark star is formed for the total intensity $|\vec{\mathcal{E}}(x, y, 0)|^2$ in the focal plane, whereby the intensity contributions $|\mathcal{E}_{x,y,z}(x, y, 0)|^2$ consist of the optical axis surrounding spots (cf. Table 4.2). Crucially, due to the added OAM the focal intensity configurations get a twist upon propagation in 3d space. This is especially visible if the first ($z = -7z_R$) and last ($z = 7z_R$) z-slice of the 3d plots in Fig. 4.4c or d are compared. As a consequence, the 3d intensity contributions $|\mathcal{E}_{x,y,z}(x, y, z)|^2$ are constituted of tilted pillars formed by the individual petals in each z-slice. Note that for a topological charge of $\ell = 2$ a similar behavior is found, whereby the twist is stronger due to the larger OAM. In contrast, if a charge of $\ell = 3$ is added (Fig. 4.4e, f), no twist is observed. Here, the sum of x- and y-contributions (f1)/(f2) form a donut in the focal transverse plane. Crucially, a significant z-contribution (f3) is created with a peak ratio of 0.83. Due to its Gaussian shape in the focal plane, it forms a cigar shape in 3d space. This shape will be of special interest for trapping applications with polarization sensitive material reacting especially to longitudinal polarization.

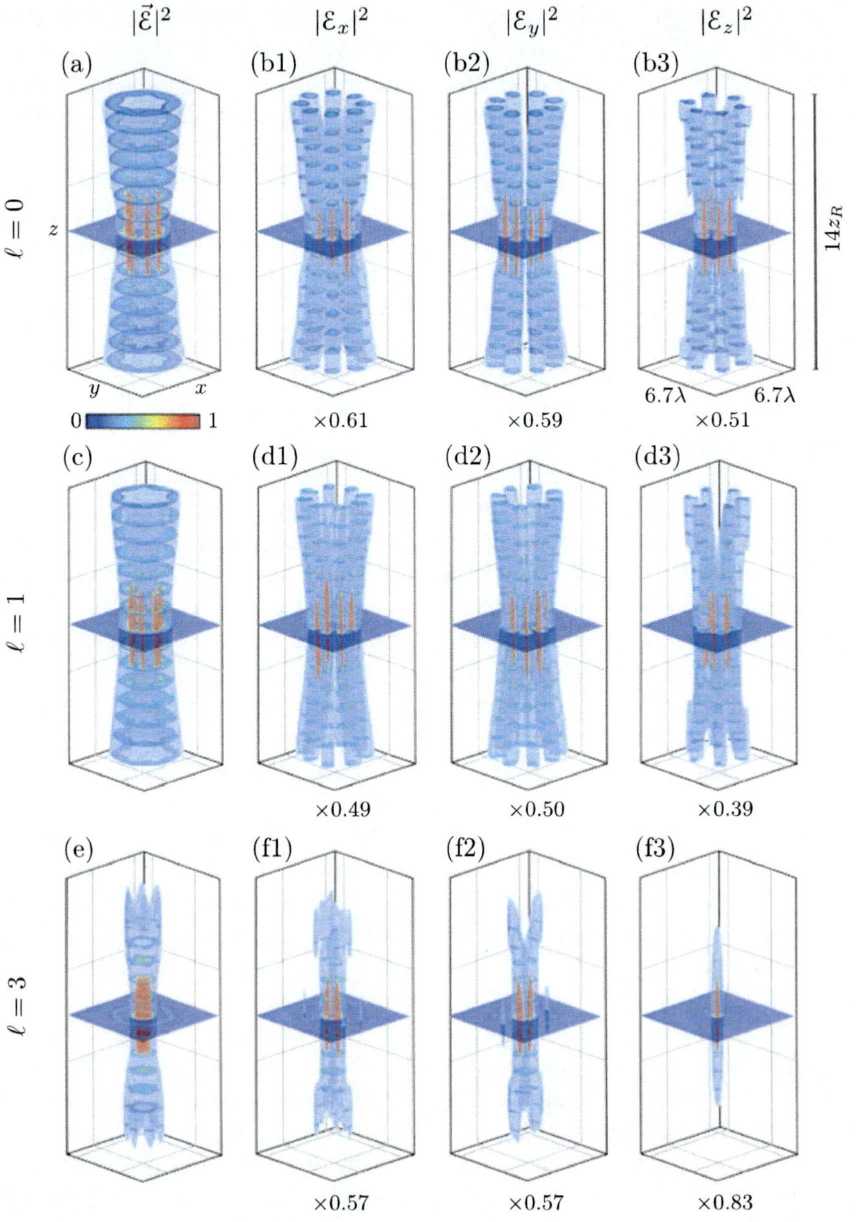

Fig. 4.4 Customization of the 3d focal volume by structuring the phase and polarization of the input light field. The **a/c/e** total normalized intensity $|\vec{\mathcal{E}}|^2$ and **b/d/f** its contributions $|\mathcal{E}_{x,y,z}|^2$ (peak ratios below) are shown originating from transverse and longitudinal electric field components. The first row corresponds to the focal structures for a pure vectorial flower ($\sigma_{12} = 8$), whereas the second and third row represent results for an incident flower combined with vortices of charge $\ell = 1$ and $\ell = 3$, respectively. Blue (red) surfaces: value of **a**, **b** 0.1/**c**–**f** 0.2 (**a**–**f** 0.8)

4.2.3 Optical Micromanipulation

Optical micromanipulation or tweezing is a topical theme in biology, medicine and many more ever since its Nobel prize awarded invention by A. Ashkin in 1970 [23, 24]. For the creation of effective optical trapping potentials for dielectric particles the strength of intensity itself as well as the spatial gradient of intensity is crucial [23–25]. Both define the overall trapping force, whereby the former corresponds to the scattering force, acting in light propagation direction, the latter to the gradient force, pointing to the location of highest intensity. By the customization of tightly focused light fields, trapping potentials can be tailored on demand to be most suitable for the trapping objects. Further, considering unconventional polarization sensitive materials, designing a 4d focal field of structured 3d polarization represent a sophisticated tool.

Due to their sophisticated intensity and polarization configurations, tailored 4d fields presented in the previous sections are especifically attractive for trapping these unconventional polarization reactive particles. To give an insight into the potential of tailored 4d fields, the exemplary trapping and orienting of zeolite-L crystals [26, 27] is demonstrated here. These nanoporous aluminosilicates are of cylindrical shape and incorporate one-dimensional nano-channels, hence, may act as nano-containers. They are able to accomodate guest molecules as drugs or polarization sensitive dyes, aligning along the nano-channels. Thus, besides their unconventional shape, loaded molecules may transform zeolites into polarization sensitive particles. This characteristic has been demonstrated for instance by applying zeolites as micro polarization sensors [28].

Here, for optical micromanipulation of these nano-containers, non-paraxial light fields created by tightly focusing a flower structure with $\sigma_{12} = 8$ and additional phase vortex of charge $\ell = \{0, 1, 2, ..., 5\}$ are applied. For the experimental realization of these 4d fields and, thus, tailored trapping potentials, the SLM-based APP (amplitude, phase and polarization) modulation system, introduced in Sects. 2.4.3 and 3.3.2, in applied in combination with a tightly focusing microscope objective (MO). The modulation system creates the input light field (wavelength $\lambda = 532$ nm), shaped in phase and polarization, which is imaged onto the back aperture of the MO. The MO (oil immersion objective, NA = 1.4) transforms the paraxial light field into the non-paraxial 4d field located in the zeolite probe (zeolite solution in water). Note that a MO of high NA is chosen as focusing objective to increase the gradient forces. This choice creates focal intensity landscapes which slightly differ from focal structures presented above, however, show very similar and, thus, perfectly appropriate characteristics for trapping zeolite-L crystals.

As first example, the pure flower configuration is focused to create a trapping potential for zeolite-L crystals. The respective trapping results are presented in Fig. 4.5. Here, (a) the transverse total intensity distribution $|\vec{\mathcal{E}}|^2$ (normalized; numerics) in the focal plane, (b) a sketch of the geometry of a zeolite-L crystal, and (c)/(d) camera images (experiment) of two trapped zeolites recorded by the inverted microscope (top view) are shown. It is well-known that, typically, cylindrical zeolite-L

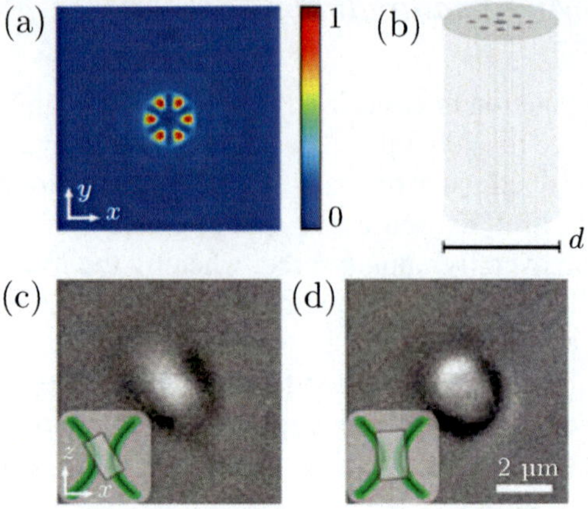

Fig. 4.5 Optically trapping zeolite-L nano-containers by a focused flower structure ($\sigma_{12} = 8$; NA $=$ 1.4, oil immersion objective). **a** Total focal intensity distribution $|\vec{\mathcal{E}}(x, y, 0)|^2$ (normalized, numerics), **b** sketch of cylindrical zeolite geometry with incorporated nano-channels, **c**/**d** camera images ((x, y)-plane) of trapped zeolite-L crystals of different basis diameter d ($d \approx 1.5\,\mu\text{m}/d \approx 2\,\mu\text{m}$). Insets indicate the orientation of zeolites in trapping potential ((x, z)-plane)

particles are trapped with its long axis approximately aligned with the optical axis of the trapping light field. In the case at hand, the trapping potential is created by the dark-stark-like intensity structure forming a tube-like 3d shape. The first trapped zeolite in Fig. 4.5c has a basis diameter d of approximately $1.5\,\mu\text{m}$, the second in Fig. 4.5d has a diameter of $d \approx 2\,\mu\text{m}$. The camera images show that the first one is trapped in a tilted position, i.e., the long axis of the zeolite is tilted compared to the optical axis, whereas the second larger one is aligned with the optical axis. This behavior is due to the fact that the size of the light field in 3d space nicely matches the larger zeolite (d), whereas the smaller zeolite (c) is tilted within the spatial structure of the field. This finding is illustrated in the insets within Fig. 4.5c and d. There, black lines indicate the envelope of the focal light field in the (x, z)-plane and green colors depict the areas of highest intensity. Zeolites are sketched in gray. The particles orients in such a way that it meets most of the high intensity regions, thus, the small particle is tilted whereas the larger one is oriented straight.

Also 4d fields formed by tightly focusing the flower structure with additional phase vortices are studied as trapping potentials of the same two zeolites. Results are given in Fig. 4.6 with (a) the total focal intensity $|\vec{\mathcal{E}}(x, y, 0)|^2$ (numerics) and camera images (experiment) of trapped zeolites ((b) $d \approx 1.5\,\mu\text{m}$, (c) $d \approx 2\,\mu\text{m}$) for the focused flower ($\sigma_{12} = 8$) with additional vortices of charge ℓ. The first column shows the same results presented in Fig. 4.5. In general, with increasing topological charge the transverse intensity structure changes similarly to results presented in Table 4.2a, i.e., its transverse diameter first decreases and subsequently increases again. Note that in 3d space the behavior for NA $= 0.9$ and NA $= 1.4$ light fields resembles as well. Naturally, trapped cylindrical particles orient according to these intensity configurations. For the larger particle, all light fields fit the particle size,

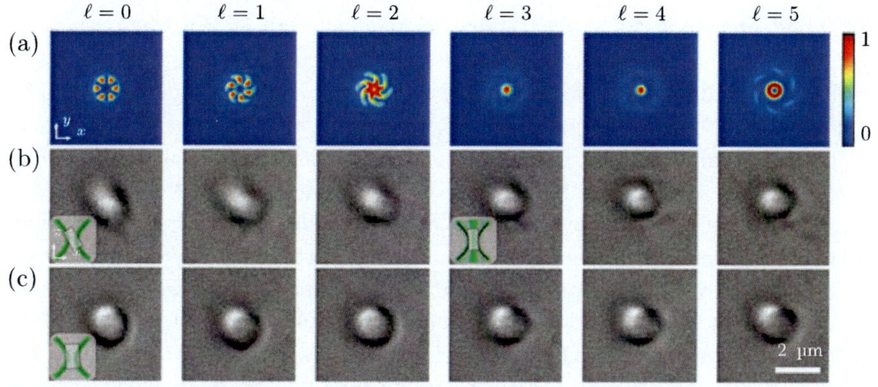

Fig. 4.6 Optically trapping and orienting zeolite-L crystals by tailored 4d fields (NA = 1.4, oil immersion objective). 4d fields are shaped by focusing a vectorial flower configuration ($\sigma_{12} = 8$) with additional phase vortices of charge ℓ. **a** Total focal intensity $|\vec{\mathcal{E}}(x, y, 0)|^2$ (normalized, numerics), and **b/c** respective camera images ((x, y)-plane) of trapped and oriented zeolite-L crystals with $d \approx 1.5\,\mu$m/$d \approx 2\,\mu$m. Exemplary insets indicate how zeolites align with trapping potential ((x, z)-plane)

thus, are oriented straight, almost parallel to the beam axis (see exemplary inset in Fig. 4.6c, $\ell = 0$). In contrast, for the smaller zeolite (Fig. 4.6b), the particle's long axis is tilted for $\ell = \{0, 1, 2\}$ with decreasing tilting angle for increasing charge ℓ. For $\ell = \{3, 4, 5\}$ the light structure matches the zeolite geometry better, thus, the particle align along the optical axis. This behavior is exemplary visualized by the inset in Fig. 4.6b, $\ell = 3$. Here the intensity is concentrated on axis. Crucially, in this case a strong longitudinal electric field contribution is found (cf. Table 4.2), which could be applied to excite polarization sensitive molecules embedded in the longitudinal nano-channels of the zeolite. Hence, besides being innovative tools for trapping and orienting zeolite-L nano-containers, 4d fields and the respective polarization configuration can be applied to selectively excite molecules accommodated within the nano-channels.

4.2.4 Conclusion

By tightly focusing tailored paraxial light fields, the on-demand shaping of non-paraxial 4d fields was evinced. Here, higher-order vector field without and with additional phase structuring were proposed as incident light field, which enable the customization of focal fields dependent on singularity indices σ_{12} and ℓ. Besides shaping transverse intensity landscapes as well as its x-, y- and z-contributions, structuring the complete focal 3d volume to form, e.g., twisted intensity configurations was demonstrated. Crucially, the ratio of transverse and longitudinal polarization components in the non-paraxial regime can be tuned, which will be of great benefit for applications including polarization sensitive materials as, e.g., 4d materials (see Sect. 4.3). As an representative example, proving the huge potential of 4d fields, zeolite-L nano-containers were optically trapped by tightly focused vectorial flower configurations with additional phase vortices. Here, the on-demand orientation of zeolite-L crystals was observed. Note that, in this case the 3d polarization of trapping potentials is of particular interest: Nano-channels of zeolites may be loaded with fluorescent dyes or pharmaceutical products, which are excited by light whose polarization is aligned along nano-channels. Hence, by the adjustment of the trapping 4d field not only the orientation of zeolites, but also the selective excitation of accommodated molecules/products is facilitated. Of course, particles of different geometry and/or polarization sensitive characteristics could be trapped as well. By adapting the singularity indices σ_{12} and ℓ of the input field the 4d field can be matched with respective geometry as well as polarization properties, thus, an optimal trapping potential can be customized.

4.3 Polarization Nano-Tomography of 4d Light Fields

As exemplified by optical trapping and orienting of zeolite-L nano-containers, tailored 4d light fields include huge potential for the advancement of applied optics. Besides contributing to the field of optical micromanipulation, the nano-scale complexity and 3d polarization nature of 4d fields is of specific interest for nano-plasmonics, material machining, high resolution imaging, or, in particular, the effective implementation of innovative 4d nano-materials.

In 4d nano-materials complex 3d material structures at the nano-scale are combined with electronic, magnetic, or optical functionalization, thus an additional degree of freedom (DoF) is added to 3d materials. Recently, functionalized nano-systems have significantly promoted the bright future of 4d nano-materials in nano-(opto)electronics, -biotechnology, or -biomedicine [29–34]. For example, stimuli-responsive nano-carriers were applied for drug delivery or peptides served as functional components in nano-systems for disease treatments [32, 35]. Most man-made nano-technology is made top-down by continuous miniaturization to the nano-scale. In contrast, in nature, the bottom-up approach is the prime strategy to construct dynamic, adaptive and learning systems at the nano-scale. This strategy includes self-assembly as an attractive route to on-demand 3d nano-structures that simultaneously exhibit an electronic, magnetic, or optical functionality [33, 34].

Prominent functionalitities are addressed optically as photo-sensitive material changes in azobenzenes. These functionalities require precise stimuli for its optimal implementation, i.e., light fields of specific characteristics exactly tailored in all three spatial dimensions as well as all its DoFs. More precisely, to optimally address 4d materials, like-wise 4d light fields structured at the nano-scale and embedding 3d polarization states are demanded [12]. As numerically demonstrated in Sect. 4.2, tightly focusing tailored paraxial fields incorporating polarization as well as phase modulation, allows for the customization of a broad range of these 4d fields. Hence, numerically the in-depth analysis of 4d fields is enabled, leading to pathbreaking findings with respect to sculpting on-demand focal fields and allowing for initial experimental applications. However, their precise experimental realization, evaluation, and application is obstructed by a lack of appropriate analysis methods. Here, the nano-scale complexity and 3d polarization nature of these non-paraxial light fields represent a major difficulty impeding the application of metrology techniques precisely working in the paraxial regime.

Therefore, there is an urgent demand for fast, i.e., single-shot, nano-tomographic techniques for the immediate identification of non-paraxial 4d fields unveiling all their DoFs—amplitude, phase as well as 3d polarization. 'Single-shot' refers to the fact that only a single measurement step, as one camera image, is needed for the 4d field identification. Hitherto, a few limited approaches have been proposed [36–38] but are failing to satisfy the topical demand: These methods are all based on slow scanning techniques with multiple shots in combination with the requirement of precise knowledge of the scanning probe characteristics and extensive reconstruction algorithms. Here, an innovative single-shot nano-tomographic approach is proposed, not requiring any data post-processing, for the identification and experimental study of focal 4d light fields by a single camera image [12]. Combining molecular self-assembly, i.e., nano-chemistry, and nano-optics, the created functional 4d nano-material itself is applied as a sensor for respective 4d light fields. By nature inspired buttom-up assembly of fluorescent rhodamine B molecules a functional molecular nano-system is formed,[4] more precisely, self-organized functionalized nano-surfaces [39, 40]. Crucially, these self-assembled monolayers (SAsMs) [41, 42] respond sensitive to amplitude, phase, as well as 3d polarization of the exciting 4d light field, enabling the qualitative single-shot analysis of complete transverse planes of the respective field at nano-scale resolution. In the following, this nano-tomographic approach is explained in detail and its functionality is proven numerically as well as experimentally by exemplary measurement of representative 4d fields.

4.3.1 *Rhodamine Self-assembled Monolayers*

To realize the single-shot nano-tomographic approach, a 4d nano-material is applied as sensor, namely, a SAsM of fluorescent sulforhodamine B silane produced on a silica glass cover slip (refractive index $n_{\text{re}} = 1.33$, thickness 170 μm). Sulforhodamine B silan has its maximal absorption at a wavelength of 572 nm and its emission maximum at 594 nm. Hence, it is a red emitting fluorophore which can be well excited by the 532 nm laser applied within experiments at hand. An idealized sketch of the

[4] The self-organized nano-surfaces or self-assembled monolayer probes are created and provided by Sebastian Lamping and (the group of) Prof. Dr. B. J. Ravoo, Organic Chemistry Institute and Center of Solf Nanoscience, University of Muenster.

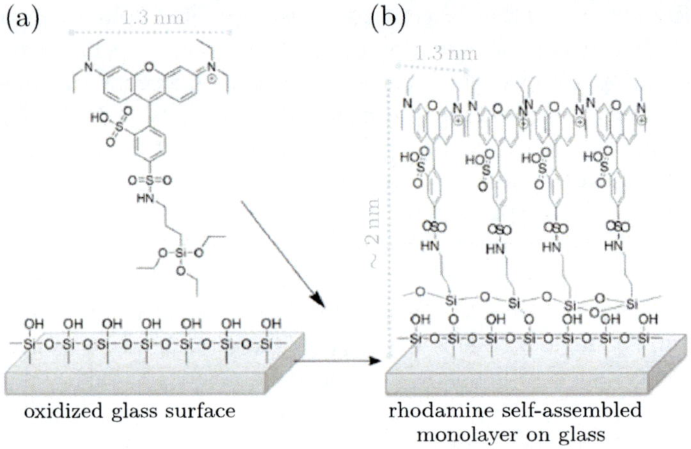

Fig. 4.7 Sketch of formation and structure of rhodamine self-assembled monolayer on glass (idealized visualization). **a** Sulforhodamine B silan is formed and combined with the oxidized glass surface to create the **b** SAsM. Dimensions are indicated in gray (dotted lines). [Adapted with permission from Ref. [12], licensed under CC BY 4.0, https://creativecommons.org/licenses/by/4.0/]

molecular SAsM structure is given in Fig. 4.7.[5] Rhodamine has a rigid structure with a broad π-system with an approximate length of 1.3 nm (a). Consider that, dependent on the tilting angle of the molecule on the glass surface (b), its transverse dimension may vary between 1.3 nm and 2 nm. Sulforhodamin B molecules interact and self-assemble in a spatially ordered way due to π-π-stacking. The respective SAsM is constituted of nanometer-sized groups of rectified molecules that occur side by side. On a micrometer level, the overall SAsM on the glass surface reveal a random orientation of its absorption dipole moments $\vec{d} = [d_x, d_y, d_z]^T$ (unit vector). Thereby, the longitudinal part d_z of dipole moments is positive due to the defined bonding site of rhodamine on silane. Note that absorption and emission spectra of rhodamine B overlap so that neighboring molecules can excite each other (self-excitation). However, the maximal interaction distance is assumed to be three to four molecules within the monolayer.

If a single molecule in a chosen (x, y)-plane is excited by the light field $\vec{\mathcal{E}}$, it reveals a fluorescence rate R_{fl} of

$$R_{\text{fl}}(x, y) = a \cdot |\vec{d} \cdot \vec{\mathcal{E}}(x, y)|^2 \tag{4.11}$$

with a representing a constant which depends on the absorption cross section and quantum yield of the molecule [36]. Considering the complete SAsM of randomly oriented molecules which is placed orthogonal to the beam's propagation direction at a chosen z-position within $\vec{\mathcal{E}}$, the fluorescence rate is calculated from [12]

[5] Details on its preparation can be found in Ref. [12].

$$R_{\text{fl}}(x, y, z) = a \cdot |\vec{d}(x, y) \cdot \vec{\mathcal{E}}(x, y, z)|^2. \tag{4.12}$$

This equation is crucial for the capability of the proposed method to identify focal 4d fields, as it will be demonstrated in the following.

4.3.2 Sensoric Properties of Monolayers

4.3.2.1 Fluorescence of Non-paraxial Fields

To demonstrate the functionality of the monolayer-based approach, i.e., the sensoric properties of rhodamine SAsM, appropriate 4d fields are required. For this purpose, the focal fields introduced in Sect. 4.2.1, Table 4.2, which are shaped by tightly focusing vectorial flower/web configurations with $\sigma_{12} = \pm 8$ and additional phase vortices of chosen charge ℓ, are applied.[6] These fields represent appropriate tools as they include amplitude, phase as well as 3d polarization structuring of well-defined shape and customizable by singularity indices σ_{12} and ℓ. Here, additional phase vortices in the input field create a distinct difference in the relative phase of each focal electric field component and clearly vary the shape as well as ratio of contributing focal components [12] (cf. Table 4.2). Numerically, 4d fields are calculated based on the approach in Sect. 4.1.2 (NA = 0.8). Experimentally, to form the respective 4d fields, the APP modulation system presented in Sect. 3.3.2 is combined with a focusing high-NA MO. As conceptionally depicted in Fig. 4.8a/b, the input field $\vec{E}_{\text{in}}$ (wavelength $\lambda = 532$ nm), formed in the plane of the SLM of the APP modulation system, is imaged onto the backaperture of the focusing MO (NA = 0.8 in air, ×100, working distance 4.5 mm)[7] to form the focal 4d field $\vec{\mathcal{E}}$. As illustrative example, the SAsM probe is located in the $z = 0$ plane of the 4d field, i.e., in the focal plane. Note that the probe is placed upside down in its holder, i.e., in beam propagation direction (bottom to top in Fig. 4.8a) the monolayer is passed before the cover glass. By this approach, abberation effects occurring when the beam is transmitted through the cover slip are avoided, as the MO focuses abberation-free in air. Hence, the undisturbed focal field can interact with the rhodamine molecules of the monolayer. The excited red fluorescence in the plane of the SAsM is observed in transmission, passing through the cover glass. For this purpose, it is collected and imaged by a second MO (NA = 0.9, ×100) and a lens onto a highly sensitive camera (Cam; Photometrics CoolSNAP MYO). To separate fluorescence from exciting light, a filter is applied in front of the camera.

---

[6]Here, a different NA is chosen with NA = 0.8 in contrast to NA = 0.9, as used for calculations of distributions within Table 4.2. This change in NA cause insignificant changes in intensity configurations and peak/mean ratio values (cf. Ref. [12]).

[7]Note that guiding mirrors need to be chosen wisely, to avoid changes within the polarization. Here, silver mirrors and a dichrotic mirror (Thorlabs DMLP605R) are incorporated within the experimental setup. To optimize the incident field at the backaperture and, thus, $\vec{\mathcal{E}}$, correcting holograms [43] are applied.

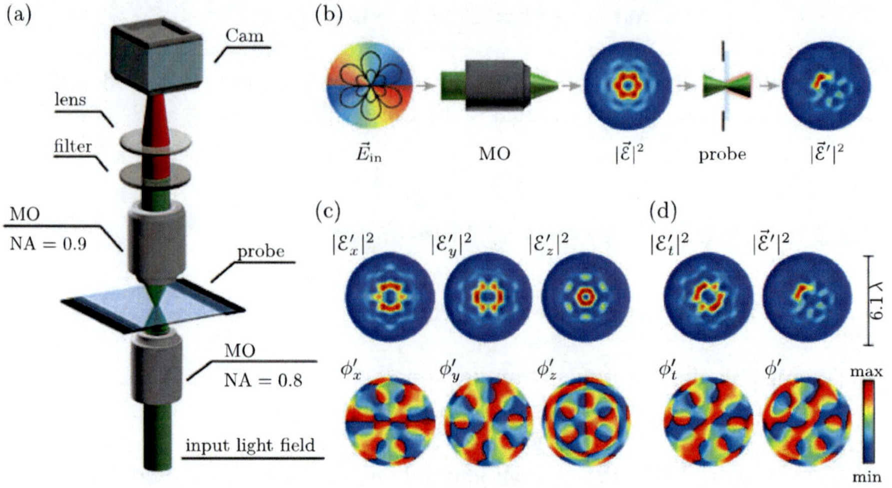

Fig. 4.8 Identification of non-paraxial 4d fields by rhodamine SAsM. **a** Sketch of the experimental system and **b** the concept for the nano-tomographic technique. **c** Intensity (normalized) and phase of the fluorescence contributions $\mathcal{E}'_{x,y,z}$ excited by the focal field components $\mathcal{E}_{x,y,z}$ and constituting **d** the transverse and total fluorescence with $\mathcal{E}'_t = \mathcal{E}'_x + \mathcal{E}'_y$ and $\vec{\mathcal{E}}' = \mathcal{E}'_t + \mathcal{E}'_z$, respectively. [Adapted with permission from Ref. [12], licensed under CC BY 4.0, https://creativecommons.org/licenses/by/4.0/]

One might expect that the random orientation of dipole moments in the SAsM may annihilate the polarization dependence of the excited fluorescence image such that fluorescence resembling a disturbed image of the total intensity $|\vec{\mathcal{E}}|^2$ in the chosen transverse plane is observed. Crucially, this is not the case [12] due to the relative phase structures $\phi_{x,y,z}$ of contributing electric field components $\mathcal{E}_{x,y,z}$ which, in addition to focal amplitude, affects the fluorescence structure significantly. This can be elucidated by numerically calculating the fluoresence rate assuming all dipoles to be oriented diagonally with $\vec{d}(x, y) = [1, 1, 1]^T/\sqrt{3}\ \forall(x, y)$ as the mean of radom orientation, giving [12]

$$R_{\text{fl}} \propto |\mathcal{E}'_x + \mathcal{E}'_y + \mathcal{E}'_z|^2. \tag{4.13}$$

Thereby, $\mathcal{E}'_{x,y,z}$ can be considered as scalar fields of the same polarization emitted by the fluorescent dipoles and inheriting the amplitude and phase of $\mathcal{E}_{x,y,z}$. As a consequence, the respective fluorescence can be seen as the interference of three scalar fields $\mathcal{E}'_x$, $\mathcal{E}'_y$, and $\mathcal{E}'_z$. Note that the choice of different dipole orientations, even varying spatially, would only change the ratio of interfering components $\mathcal{E}'_{x,y,z}$ in Eq. (4.13).

To study this effect, the fluorescence at $z = 0$ excited by the tightly focused (NA = 0.8) exemplary input field $\vec{E}_{\text{in}}$ with $\sigma_{12} = 8$ and $\ell = 2$ is numerically calculated for $\vec{d} = [1, 1, 1]/\sqrt{3}$. More precisely, the field $\vec{\mathcal{E}}' = \vec{e} \cdot [\mathcal{E}'_x + \mathcal{E}'_y + \mathcal{E}'_z]$ ($\vec{e}$:

normalized polarization vector of emitted field)[8] is determined, with results presented in Fig. 4.8c, d. Comparing the intensity of the scalar fields $\mathcal{E}'_{x,y,z}$ to the one of the focal field contributions $\mathcal{E}_{x,y,z}$ in Table 4.2a, they clearly resemble. The shown phase $\phi'_{x,y,z}$ is also equal to the relative phases of the focal field $\phi_{x,y,z}$. The phases of the transverse components $\phi'_{x,y}$ resemble each other with respect to their structure only differing in angular orientation. Their structure include a central double-charged vortex (/two very close singly charged vortices), whereas the phase of the longitudinal contribution ϕ'_z only embeds a singly charged vortex close to the optical axis. This fact has a significant influence on the resulting fluorescence image: While the sum of transverse components $\mathcal{E}'_t = \mathcal{E}'_x + \mathcal{E}'_y$, visualized in Fig. 4.8d, reveal a symmetric structure keeping the double-charged vortex (/two very close singly charged vortices), the total observed fluorescence image $|\vec{\mathcal{E}}'|^2$ shows an asymmetric structure. This asymmetry originates from the spatially varying phase difference between $\mathcal{E}'_t$ and $\mathcal{E}'_z$, resulting in constructive interference top left from the optical axis and destructive at the bottom right within the transverse plane. Hence, in contrast to the expectation of observing a disturbed image of the focal intensity $|\vec{\mathcal{E}}|^2$, an asymmetric intensity distribution $|\vec{\mathcal{E}}'|^2$ is found (cf. concept in Fig. 4.8b). Crucially, the observed fluorescence image is characteristic for the exciting focal 4d field, only appearing due to the interaction of transverse and non-negligible longitudinal field components including their amplitude and phase. Inversely, the asymmetric shape clearly reveals the significant contribution of non-paraxial z-components and, thus, the 3d polarization. Hence, the method is capable of identifying the fourth dimension of 4d fields, which is invisible to paraxial measurement techniques. Additionally, the SAsM enables the single-shot detection of the distinctive fluorescence image and, thereby, the spatially resolved identification of the typically invisible z-polarization without any data post-processing. Moreover, the spatial resolution of measurement results is at the nano-scale in all three spatial dimensions: Transverse resolution is limited by the size of molecules (see above) in combination with the optical imaging system; longitudinally, the thickness of the monolayer, thus the molecular nano-scale, defines resolution.

4.3.2.2 Fluorescence of Paraxial Fields

As the asymmetric fluorescence shape originates from the significant contribution of longitudinal exciting field components, the nano-tomographic approach can clearly differentiate between non-paraxial and paraxial light fields. Paraxial fields only embed negligable z-components, so that, in contrast to non-paraxial fields, the symmetry of the fluorescence will adapt to the symmetry of the exciting field. To demonstrate this effect, the fluorescence image, excited by the paraxial version of a focused higher-order vector beam with $\sigma_{12} = \pm 8$ and additional phase vortex $\ell = \{0, 1, 2, ..., 5\}$, is numerically calculated. For this purpose, the exciting light

---

[8]Note that this field is also called fluorescence (field) in the following, as it only varies by a constant factor from the real fluorescence rate and normalized images are shown.

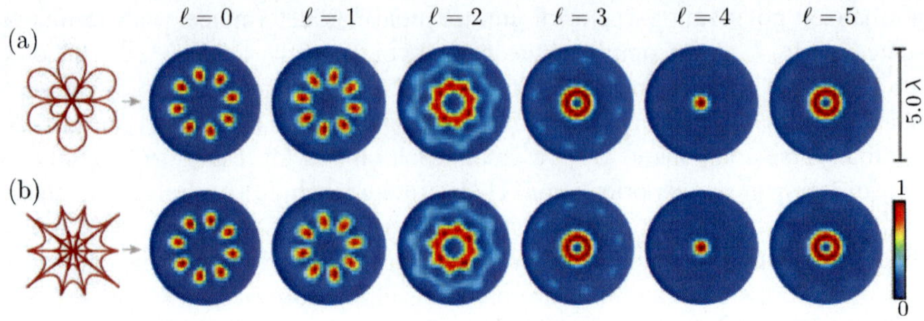

Fig. 4.9 Monolayer fluorescence excited by paraxial light fields, namely, the far-field distributions of a vectorial **a** flower or **b**) web structure with additional phase vortices of charge ℓ. Intensity images are normalized and resolution adapted to the following experiment. [Adapted with permission from Ref. [12], licensed under CC BY 4.0, https://creativecommons.org/licenses/by/4.0/]

field is assumed to be the far-field distribution of the incident vector beams $\vec{E}_{\text{in}}$, i.e., $\vec{\mathcal{E}}_{\text{far}} = \mathcal{F}(\vec{E}_{\text{in}}(x,\ y,\ 0))$ [12]. Within numerics, the SAsM is now chosen to be of random orientation $\vec{d}(x,\ y)$ positioned at $z = 0$. By Eq. (4.12) with $\vec{\mathcal{E}} = \vec{\mathcal{E}}_{\text{far}}$ and $\mathcal{E}_z \approx 0$, the normalized fluorescence distribution is determined. Results are presented in Fig. 4.9 for the exciting far-field of the (a) flower and (b) web configuration with additional phase vortices. Note that, here and in the following section, for matching experiments (Sect. 4.3.3) and numerics, the SAsM is assumed as a matrix of dipole vectors (random orientation), whereby some neighboring vectors are of the same orientation, considering group building (see Sect. 4.3.1). Beyond, the ratio of respective matrix pixel size and the one of the numerical electric field is adjusted in such a way that the dimension of molecules and exciting light field matches [12]. Further, after calculation, the resolution of fluorescence structure is decreased in order to adapt numerical results to the resolution of the experimental system, limited by the MO, lens and camera.

Due to its paraxial characteristics with no z-contribution, the fluorescence images are all of axial symmetric shape, following the cylindrical symmetry of exciting fields [12]. Comparing the results for the flower and web structures (Fig. 4.9), there are only slight differences visible, even though the respective input polarization structures are significantly different. In contrast, for non-paraxial fields, not only the differentiation between focused flower and web configurations is enabled, but also distinctive fluorescence structures dependent on σ_{12} and ℓ are observed, as it will be evinced in the next section.

4.3.3 Experimental Identification of Focal Fields

To observe the distinctive fluorescence structures for non-paraxial 4d fields, respective focused flower and web configurations with additional phase vortices are experi-

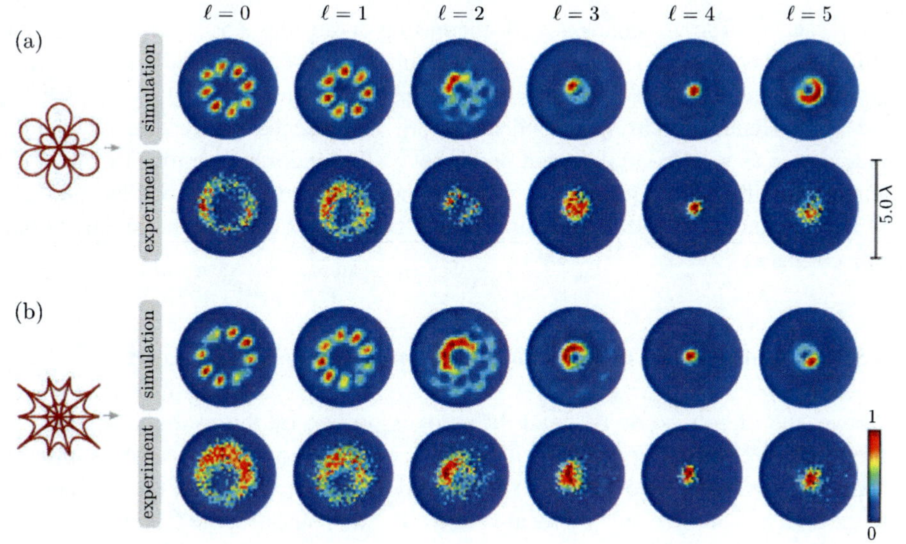

Fig. 4.10 Experimental identification of 4d light fields in the focal plane. The numerically calculated (top) and experimentally measured (bottom) normalized fluorescence structures are shown for a tightly focused (NA $= 0.8$) vectorial **a** flower and **b** web configuration ($\sigma_{12} = \pm 8$) with additional phase vortices of charge ℓ. Resolution of simulations is adapted to experimental constraints. [Adapted with permission from Ref. [12], licensed under CC BY 4.0, https://creativecommons.org/licenses/by/4.0/]

mentally created and radiated on the monolayer probe [12] as explained in Sect. 4.3.2 (cf. Fig. 4.8a, b). Respective numerical (top) as well as experimental (bottom) results are presented in Fig. 4.10a, b for the incident flower and web, respectively. All intensity images are normalized to its own maximum. To be more accurate, the noise within recorded camera images is reduced by background subtraction and taking the mean value of ten images.[9]

As characteristic for the tomographic approach, for both, focused flower and web, asymmetric intensity distributions are observed within simulations, as especially visible for $\ell \geq 2$. These numerics are perfectly confirmed by experimental results (see Ref. [12] for detailed comparison). Note that experimental deviations from numerics originate, for instance, from non-exact perpendicular orientation of the probe in relation to the optical axis. Further, in particular, the limited resolution of the detecting imaging system in combination with low monolayer fluorescence power (compared to background noise) and the self-excitation of molecules needs to be considered. Crucially, the detected asymmetry distinctly proves the significant contribution of non-paraxial z-components and, thus, confirms the pathbreaking experimental detection of typically invisible non-paraxial field properties [12]. Following the characteristics of the exciting focal fields (cf. Table 4.2), the transverse diameter of the

[9]Note that this optimization is easily implemented by appropriate camera settings, thus, no additional post-measurement steps are required.

fluorescence structure changes with increasing ℓ. Besides observed asymmetry in simulation and experiment, by presented results (Fig. 4.10), focused flower and web configurations are clearly distinguishable, as e.g. by differences within the asymmetry for $\ell = 0$. Hence, the strong dependence of the fluorescence structure on focal amplitude, phase as well as 3d polarization is proven. In total, results prove the method's single shot ability to qualitatively visualize non-paraxial field properties and, thus, to identify and study focal 4d fields without the need of any post-measurement data processing.

4.3.3.1 Outlook on Full 4d Nano-Tomography and Field Reconstruction

Above, the identification of focal 4d fields is proven by the example of applying the monolayer-based approach within the focal plane at $z = 0$. Nevertheless, the approach can easily be extended to the full 4d nano-tomography of the 4d field, i.e., to the analysis of the whole 3d focal volume. For this purpose, the monolayer will be moved through the volume scanning different z-slices. As the SAsM is of nanometer thickness, a longitudinal resolution at the nano-scale is enabled and, hence, the whole 4d field can be analyzed with nanometer resolution in all three spatial dimensions with a single-shot per z-slice.

Beyond, advancing the monolayer approach, the full reconstruction of 4d electric fields is facilitated [12]. Here, molecular layers of defined, pure x- ($\vec{d} = [1, 0, 0]^T$) and y- ($\vec{d} = [0, 1, 0]^T$), as well as diagonal orientation ($\vec{d} = [1, 1, 1]^T/\sqrt{3}$) are required ($\forall\,(x, y)$). For this purpose, to realize an even improved system, quinacridone [44] could be applied, as it shows stronger fluorescence and less self-excitation for the exciting wavelength. As demanded for full reconstruction, this fluorescent molecule can also be arranged in highly-order 2d structures by programmable self-assembly [12]. By these innovative SAsM the single-shot detection of each transverse electric field components separately, including intensity and phase, will be enabled, as the fluorescence is proportional to $\vec{d} \cdot \vec{\mathcal{E}}$. Knowing these components, the detected fluorescence excited in the case of diagonally oriented monolayers can be applied to caculate the longitudinal field contribution.[10] Crucially, combining this advanced approach with the z-scanning of the 3d volume, the full reconstruction of the whole 4d field in 3d space is facilitated by a very efficient, fast technique at nano-scale resolution.

[10]Note that the fluorescence excited in the case of $\vec{d} = [0, 0, 1]$, $\forall(x, y)$ would not provide the desired results, as the respective fluorescence polarization (mainly oriented longitudinally) could not be imaged by the applied imaging system.

4.3.4 Conclusion

Recently, a broad range of applications as the optimal implementation of 4d nanomaterials are awaiting the inclusion of tailored 4d light fields. However, the nanoscale complexity as 3d polarization nature impede its in-depth experimental evaluation and, thus, application. Here, a method was proposed and experimentally demonstrated to solve this issue by a single-shot nano-tomographic approach not requiring any data post-processing, but enabling the identification of typically invisible non-paraxial field properties [12]. The approach is based on the application of a functionalized nano-surface, namely, self-assembled monolayer of sulphorhodamine B molecules. Due to its molecular structure, the SAsM facilitates the analysis with nano-scale resolution in all three spatial dimensions. The fluorescent monolayer reacts amplitude, phase as well as 3d polarization sensitive, resulting in distinctive fluorescence structures for 4d fields. This characteristic was elucidated numerically and clearly confirmed experimentally by the example of 4d fields formed by tightly focusing flower and web configurations with additional vortices. Additionally, as presented above, the method can be extended to access and fully reconstruct the whole 3d volume of focal 4d electric fields. In total, the proposed approach finally presents the demanded nano-technological tool for the identification and study of 4d fields and, thus, facilitates the pending advancement in applied optics, where 4d fields enable, e.g., the effective implementation of 4d materials.

4.4 Optical Singularities and 3d Topological Structures

Besides being of interest for shaping sophisticated 4d light fields with complex intensity configurations and 3d polarization (cf. Sects. 4.2, 4.3), tightly focusing tailored singular vector fields also enables the customization of 3d polarization singularities in non-paraxial fields. The study of these is fundamentally meaningful giving new insights into the not yet fully understood nature of light. Furthermore, the crucial z-polarization components are responsible for the formation of complex 3d topological structures in 3d space as optical cones, twisted ribbons or Möbius strips. Typically, these well defined structures are formed by 3d polarization ellipses around generic polarization singularities in non-paraxial fields [9]. In these fields, C-points of undefined orientation and L-points of undefined handedness represent generic point singularities in 2d planes, which form line singularities meandering throughout 3d space. Until now, first examples of complex topological structures have been demonstrated as the creation of single- or multi-half-twist Möbius strips around an individual on-axis singular point [10, 45]. However, sculpting topological structures has not reached it full potential yet as shaping 3d polarization singularities has not been considered as customization tool. In this section, by numerical calculations, tailored singularities and not yet known intriguing topologies are presented [13], formed by tightly focusing ($\mathrm{NA} = 0.9$) higher-order vector beams with additional phase vortices (cf. Table 4.2, Eq. (4.10)). As evident from results in Sect. 4.2.1, an incident pure vector fields of index σ_{12} cause the formation of focal states of polarization whose major axis is oriented parallel or perpendicular to the focal plane ($z = 0$). By the application of an additional phase vortex of topological charge ℓ, the symmetry between orthogonal polarization components of the incident field $\vec{E}_{\text{in}}$ is broken enabling tailored topologies: A higher-order vector field can be represented as a superposition of two scalar light fields of opposite helical phase structure (OAM) and orthogonal circular polarization (cf. Sect. 2.3.2, Eq. (2.26)). Hence, by the addition of a global phase vortex, the symmetry between polarization components is broken, resulting in a dramatic change in focal generic singularities and topological structure centered on- and off-axis. In the following, the control of generic C- and L-singularities in the focal plane ($z = 0$) by the adaption of σ_{12} and ℓ is demonstrated as well as the on-demand structuring of topological structures, even enabling innovative optical Möbius strip arrays in non-paraxial fields [13]. Besides contributing significantly to the fundamental study of singular optics, presented results are of specific interest for optical fabrication of novel functional media base on non-trivial topologies or the assembly of polarization sensitive particles.

4.4.1 Generic Singularities in Non-paraxial 4d Fields

4.4.1.1 Analysis Methods for 3d Polarization Singularities

In the paraxial regime, Stokes parameters or, more precisely, complex Stokes fields serve as a tool for the identification and characterization of polarization singularities. Due to the 3d polarization nature, this analysis technique is not applicable in the non-paraxial regime. Here, 3d polarization and the overall topology of the 4d field can be deciphered by the major axis $\vec{p}_0$, minor axis $\vec{q}_0$, and normal vector $\vec{N}^{\text{p}}$ of respective 3d polarization ellipses (cf. Sect. 2.3.1). For the non-paraxial field $\vec{\mathcal{E}}$, these are defined according to [46]

$$\vec{p}_0 = \frac{1}{|\sqrt{\vec{\mathcal{E}}\vec{\mathcal{E}}}|}\Re\left(\vec{\mathcal{E}}\sqrt{\vec{\mathcal{E}}^*\vec{\mathcal{E}}^*}\right), \tag{4.14}$$

$$\vec{q}_0 = \frac{1}{|\sqrt{\vec{\mathcal{E}}\vec{\mathcal{E}}}|}\Im\left(\vec{\mathcal{E}}\sqrt{\vec{\mathcal{E}}^*\vec{\mathcal{E}}^*}\right), \tag{4.15}$$

$$\vec{N}^{\text{p}} = \Im\left(\vec{\mathcal{E}}^* \times \vec{\mathcal{E}}\right) = 2\vec{p}_0 \times \vec{q}_0. \tag{4.16}$$

Note that $\vec{p}_0$ and $\vec{q}_0$ do not have a defined direction (see square root), thus, represent headless vectors, which matches the π-invariance of polarization ellipses under rotation [47]. In contrast, $\vec{N}^{\text{p}}$ is a vector of defined direction, which defines the polarization states handedness (normal vector of oscillation plane $\vec{n}_{\text{osc}} = 2[|p_{0,x}|, |p_{0,y}|]^T \times [|q_{0,x}|, |q_{0,y}|]^T$; $\vec{N}^{\text{p}} \parallel \vec{n}_{\text{osc}}$: right-handed, $-\vec{N}^{\text{p}} \parallel \vec{n}_{\text{osc}}$: left-handed) [48].

For the occurrence of a C-singularity, major and minor axis of respective polarization state(s) need to be of the same length, i.e., $p_0 = q_0$ ($|\vec{p}_0| = p_0$, $|\vec{q}_0| = q_0$). To identify respective singularities, one may apply the complex scalar field [46]

$$\Psi_C = \vec{\mathcal{E}}\vec{\mathcal{E}} = \left(p_0^2 - q_0^2\right) \mathrm{e}^{\mathrm{i}2\gamma} \tag{4.17}$$

with $\vec{p}_0\vec{q}_0 = 0$ and rectifying phase γ (cf. Eq. (2.19)). Hence, for a C-singularity with $p_0 = q_0$ the scalar field is $\Psi_C = 0$. Therefore, C-singularities can be identified by intersections of zero-lines (-surfaces) $Z_{\Re}$ and $Z_{\Im}$ with $\Re(\Psi_C) = 0$ and $\Im(\Psi_C) = 0$, respectively, in 2d (3d) space. The generic form of C-singularities in non-paraxial fields is a point in 2d and line in 3d space [49, 49–51]. This is due to the codimension two of this singularity: for their occurrence non-rectified vector $\vec{p}$ and $\vec{q}$ (cf. Sect. 2.3.1) need to be perpendicular and of the same length.

For the identification of L-singularities in the non-paraxial regime, the normal vector $\vec{N}^{\mathrm{p}}$ is applied. Here, linear states of polarization are searched, i.e., the minor axis needs to vanish. This requirement is equivalent to $\vec{N}^{\mathrm{p}} = [N_x^{\mathrm{p}}, N_y^{\mathrm{p}}, N_z^{\mathrm{p}}]^T = \vec{0}$ [46]. Consequently, for identifying L-singularities, zero-lines in 2d (-surfaces in 3d) space defined as $N_x^{\mathrm{p}} = 0$, $N_y^{\mathrm{p}} = 0$, and $N_z^{\mathrm{p}} = 0$ are studied. If all three lines (surfaces) cross in a joint point (line), a L-singularity is found. To determine the generic form of L-singularities, the non-rectified vectors $\vec{p}$ and $\vec{q}$ are considered. For a linear state, $\vec{p} \parallel \vec{q}$ needs to be fulfilled, i.e., both polar angles of these vectors have to be equal [52], revealing a codimension of two. Thus, in contrast to paraxial L-singularities with a codimension of one, the generic form of a non-paraxial L-singularity represents a point in 2d and line in 3d space [46, 53, 54].

4.4.1.2 Customization of Non-paraxial Singularities

In order to get an expressive insight into the ability to customize 3d polarization singularities in 4d fields, here, the generic point singularities in the focal 2d plane ($z = 0$) of tailored 4d fields are investigated. As elucidated above, tightly focusing vector beams of higher order with additional phase vortex represent an appropriate tool for sculpting the topology and singularities in 4d fields [13]. As illustrative example, a vectorial flower with $\sigma_{12} = 8$ is chosen as input field $\vec{E}_{\mathrm{in}}$ with vortices of charge $\ell = \{0, 1, 2, ..., 5\}$ corresponding to the non-paraxial light fields already presented in Table 4.2a. Within its non-zero focal intensity areas $|\vec{\mathcal{E}}(x, y, z = 0)|^2 \neq 0$ these numerically calculated focal fields are examined for C- and L-singularities by zero-lines $Z_{\Re,\Im}$ and $N_{x,y,z}^{\mathrm{p}} = 0$, respectively. Respective results are presented in Fig. 4.11. Here, the total focal intensity $|\vec{\mathcal{E}}(x, y, 0)|^2$ for each ℓ is illustrated in the background with zero-lines on top ((a) $Z_{\Re,\Im}$: blue, orange; (b) $N_{x,y,z}^{\mathrm{p}}$: red, yellow, green) (Fig. 4.11).

For the focused pure vector beam with $\ell = 0$, neither a generic C- nor a L-singularity is detected. In Fig. 4.11a, zero-lines $Z_{\Re}$ and $Z_{\Im}$ exactly overlap, i.e., lines of pure circular polarization are found. However, C-lines do not represent generic

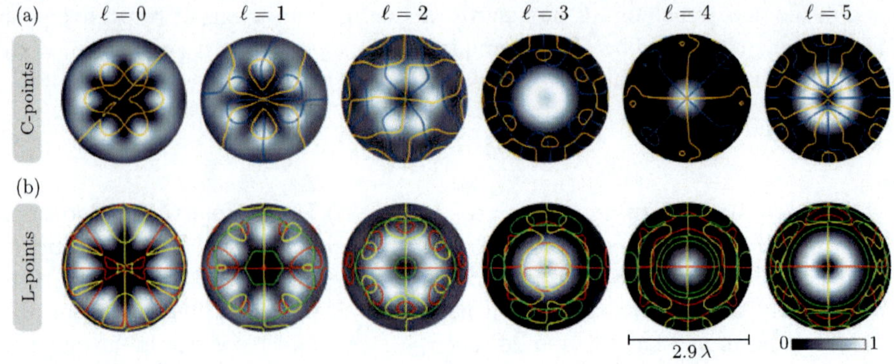

Fig. 4.11 Customization of 3d polarization singularities in the non-paraxial regime by tightly focusing (NA $= 0.9$) a vectorial flower ($\sigma_{12} = 8$) with additional phase vortices of charge ℓ. Shown are zero-lines **a** $Z_{\Re,\Im}$ (blue, orange) and **b** $N^{\mathrm{p}}_{x,y,z} = 0$ (red, yellow, green) on the total normalized intensity $|\vec{\mathcal{E}}(x,\ y,\ 0)|^2$ for the identification of C- and L-points, respectively. [Adapted with permission from Ref. [13], licensed under CC BY 4.0, https://creativecommons.org/licenses/by/4.0/]

singularities in the 2d focal plane. Further, in Fig. 4.11b, there are no points found in which all three zero-lines of $N^{\mathrm{p}}_{x,y,z}$ intersect, hence, no L-points are detected. In contrast, by increasing the topological charge ℓ, generic singularity can be created within the focal plane. As evident from zero-line intersections, for $\ell = 1$ and $\ell = 2$ the 4d field embeds a singularity network (SN) of (a) six C- and (b) six L-points in its focal plane within regions of $|\vec{\mathcal{E}}|^2 \neq 0$. Note that between two L-points there is always a C-point found and vice versa. For $\ell = 3$, C-points are vanishing, i.e., no C-point is observed, whereas an individual on-axis L-singularity is identified. Inversely, for $\ell = 4$ no L- but an on-axis C-point is found. For the latter, the C-point is marked by the intersection of two $Z_\Re$ and two $Z_\Im$ zero-lines. Hence, here, two very close generic C-points are highlighted or a single non-generic higher-order C-point is located. As the latter is unstable under minor perturbations, its unfolding would be expected within, e.g., an experimental realization. Increasing the topological charge to $\ell = 5$, the occurrence of two C-points and no L-points within the donut intensity configuration is observed.

In total, the adaptability of singular properties is evinced within the focal light field by adjusting the incident field. By additional numerical studies with $\sigma_{12} \in [4,\ 12]$ and $\ell \in [0,\ 7]$ one can derive the following rules [13]

I)	$\ell = 0$	: no generic singularity,
II)	$0 < \ell < \vert\sigma_{12}\vert/2 - 1$	: $\vert\sigma_{12} - 2\vert$ C- and L-points,
III)	$\ell = \vert\sigma_{12}/2 - 1\vert$	: one on-axis L-point,
IV)	$\ell = \vert\sigma_{12}\vert/2$	: one on-axis C-point.

Rule II) corresponds to the creation of a non-paraxial SN within the focal plane, whereby the number of focal point singularities is related to the number of incident

ζ-lines $|\sigma_{12} - 2|$. Clearly, the focal field and its singularities can be tailored by the choice of index σ_{12} and ℓ, as confirmed by all rules.

4.4.2 From Cone to Möbius Strip Arrays

Besides controlling the polarization singularities embedded within the 4d fields, of course, the surrounding polarization structure and, thus, the optical topology is affected as well by the adaption of the incident paraxial field and its indices σ_{12} and ℓ [13]. To study the light field topology, an elegant method is given by the calculation of the major axis, minor axis or normal vector of 3d polarization states at all points on a circle in the focal plane [10, 50, 54]. By this approach, optical cones, ribbons or Möbius strips can be identified. In the following, on the one hand, the topology around the optical axis of an exemplary 4d field is numerically studied. Additionally, on the other hand, the topological structure around off-axis generic singularities is investigated by numerical calculations, leading to the finding of not yet observed complex polarization topologies.

4.4.2.1 Topologies Around the Optical Axis

As representative example [13], the incident vectorial flower configuration with $\sigma_{12} = 8$ and $\ell = 1$ is chosen, being the simplest example of breaking the symmetry by an additional vortex. As visualized in Fig. 4.12, (a) the total intensity structure $|\mathcal{E}|^2 \in [0,\ 1]$ in the focal plane resembles a deformed dark star, whereby (b) intensity contributions $|\mathcal{E}_{x,y}|^2 \in [0,\ 1]$ and $|\mathcal{E}_z|^2 \in [0,\ 1]$ represent configurations of $|\sigma_{12}|$ and $|\sigma_{12} - 2|$ petals around the optical axis. All intensity structures are normalized to their respective maximum with its peak ratio shown within the image. Obviously, z-components contribute significantly in this case (peak ratio of $\times 0.39$). Note that for $\ell = 0$ similar distributions are achieved, however, the key for the realization of complex topological structures is hidden within respective phase distributions. For $\ell = 0$, discrete phase structures are realized (cf. Fig. 4.3a). In contrast, phase distributions $\phi_{x,y,z} \in [0,\ 2\pi]$ for $\ell = 1$ embed crucial phase vortices, as illustrated in the top right of each intensity image in Fig. 4.12b. These cause the appearance of sophisticated topologies around the optical axis, as demonstrated in Fig. 4.12c–e.

The polarization topology on circles around the optical axis is determined for three different radii, as indicated in Fig. 4.12a by white dashed circles. For all radii, i.e., for (c) the smallest, (d) the medium, and (e) largest circle, the major axis (blue lines with blue and green endpoints; unit length) of polarization ellipses on each circle (black) is traced. The projections of major axis, i.e., of the respective polarization topology, is shown in the $(x,\ y)$-plane of each plot. Interestingly, the formed electric field and phase distributions yield different distinct topological configurations for each circle. For the (c) smallest circle, a cone-like topology created by major axes is found. On this circle, longitudinal polarization components clearly dominate in comparison to

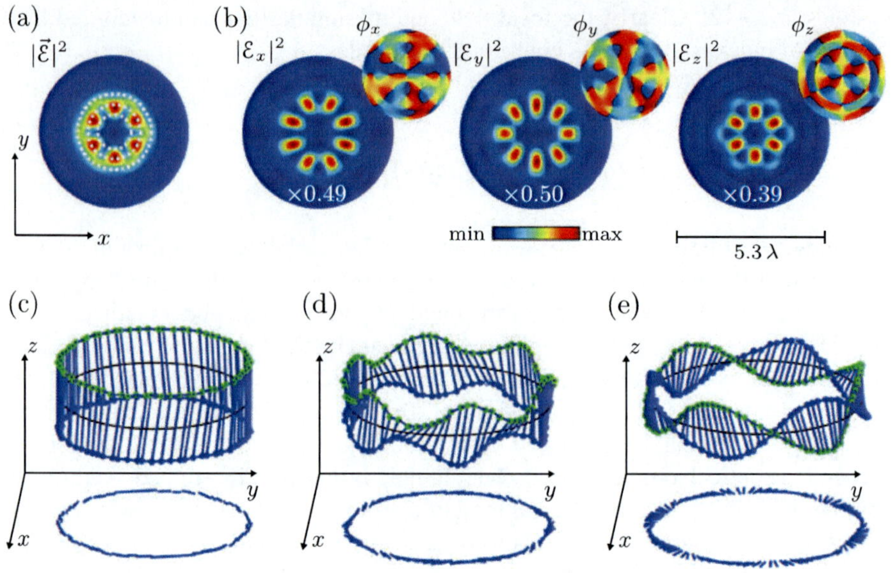

Fig. 4.12 From cone to twisted ribbon. Tailored 3d polarization topologies around the optical axis in the focal plane ($z = 0$) formed by tightly focusing (NA $= 0.9$) a vectorial flower configuration with $\sigma_{12} = 8$ and $\ell = 1$. **a** Total focal intensity $|\vec{\mathcal{E}}|^2 \in [0,\ 1]$ and **b** intensity contributions $|\mathcal{E}_{x,y,z}|^2 \in [0,\ 1]$ with respective relative phase structures $\phi_{x,y,z} \in [0,\ 2\pi]$ are shown. 3d topological structures are created by tracing the major axis of polarization ellipses (blue lines with blue and green endpoints) located on the **c** smallest, **d** medium, and **e** largest circle, highlighted in **a** by white dashed line. Projections of major axes are given in $(x,\ y)$-plane of each plot in **c**–**e**. [Adapted with permission from Ref. [13], licensed under CC BY 4.0, https://creativecommons.org/licenses/by/4.0/]

transverse ones, resulting in the cone arrangement. If (d) the circle radius is increased, the contribution of transverse components raises, causing a transverse wiggling of major axes on the circle. Hence, the cone structure in Fig. 4.12c degenerates into a ribbon without full twists in Fig. 4.12d for the medium circle. Full twists are avoided due to the, in total, still dominant contribution of z-components. In contrast, if (e) the largest circle is considered, this overall dominance is erased, enabling the formation of a twisted ribbon topology around the optical axis. Six twists are observed, corresponding to the number of intensity lobes in $|\mathcal{E}_z|^2$ and, thus, number of ζ-lines $|\sigma_{12} - 2|$ of the incident light field.

An equivalent topology analysis was performed for incident fields with $\sigma_{12} \in [4,\ 10]$ and $\ell \in [0,\ 7]$. For a set singularity index σ_{12} and charge $0 < \ell < |\sigma_{12}|/2 - 1$, a change from cone, via distorted cone, through to ribbon with $|\sigma_{12} - 2|$ twists is observed for increasing circle radius. For larger topological charges $\ell \geq |\sigma_{12}|/2 - 1$, point- or donut-shaped intensity structures are realized, as demonstrated in Table 4.2a. In these cases, cone topologies or ring configurations (all major axis lay in focal plane) are formed by major axes. Hence, in total, by tightly focusing these vectorial fields, intriguing polarization topologies are created around the optical axis, which are customizable by the adaption of singularity indices σ_{12} and ℓ [13].

4.4.2.2 Off-Axis Topologies

Typically, 3d polarization topologies are investigated around generic polarization singularities as evident from previous work [50, 54]. In this case, specific indices can be defined for its characterization. Following this approach, here, topological structures around generic polarization singularities are explored numerically [13] for the same tailored focal 4d field as studied above ($z = 0$), i.e., $\vec{E}_{\text{in}}$ with $\sigma_{12} = 8$ and $\ell = 1$. Consider that, in contrast to typically investigated fields, here, generic C- and L- polarization singularities (six of each) are customized to be located off axis in an extended SN, whereby the study at hand concentrates on off-axis C-points and surrounding major axes $\vec{p}_0$.

To fully identify topologies on a circle κ around generic 3d polarization singularities, projections on three different planes are applied [9, 50, 54]. The first projection of major polarization axes is performed onto the (x, y)-plane, also called Σ_0-plane. For this projection plane, two indices I and Λ are defined [49, 51] with I describing the winding number of projected axes for one cycle around the considered circle. Index Λ gives the number of ζ-lines created by the projections. Additionally, major axes are projected onto the so-called radial plane τ_0 and tangential plane π_0 [9, 50, 54]. Respective planes are visualized in Fig. 4.13a. Here, the major axis $\vec{p}_0$ (blue) on the circle κ (gray) at angular position $\rho \in [0, 2\pi]$ is projected on (a1) τ_0 and (a2) π_0 (red), forming projections $\vec{T}$ and $\vec{\Pi}$ (green with red endpoint), respectively. For each plane, an index is defined with

$$\tau_p = \frac{\delta(2\pi) - \delta(0)}{2\pi}, \qquad \pi_p = \frac{\delta(2\pi) - \delta(0)}{2\pi}, \tag{4.18}$$

representing the winding number of projections $\vec{T}$ and $\vec{\Pi}$, respectively, after a 2π-cycle around circle κ. Thereby, $\delta(\rho)$ corresponds to the angle between $\vec{T}$ or $\vec{\Pi}$ and the x'- or y'-axis (cf. Fig. 4.13(a1) or (a2)), respectively.

Similar to singularity indices, which are defined for paraxial light fields describing, e.g., embedded generic singularities and surrounding polarization structures, characteristic indices and, thus, topologies are expected for generic C-points in non-paraxial fields [50]. Considering a generic C-point laying in the Σ_0-plane ($\vec{N}^{\text{p}} \perp \Sigma_0$),[11] traced major axis typically generate Möbius strips with a single or three half-twists [50]. In these cases, projection onto the Σ_0-plane reveals an index of $I = \pm 1$ and $\Lambda = 1$ or 3. Reflecting the number of twists, projections onto τ_0 and π_0 give characteristic

[11]Note that characteristic topologies also appear for C-point not laying in the (x, y)-plane (e.g., Ref. [45]). In this case, the plane of interest in which κ is drawn and which is applied for the determination of I and Λ, corresponds to plane perpendicular to the C-point's normal axis $\vec{N}^{\text{p}}$ [50]. However, in the study at hand (x, y)-plane is always chosen for the respective analyses, as C-points' tilting angles with respect to this plane is negelectable.

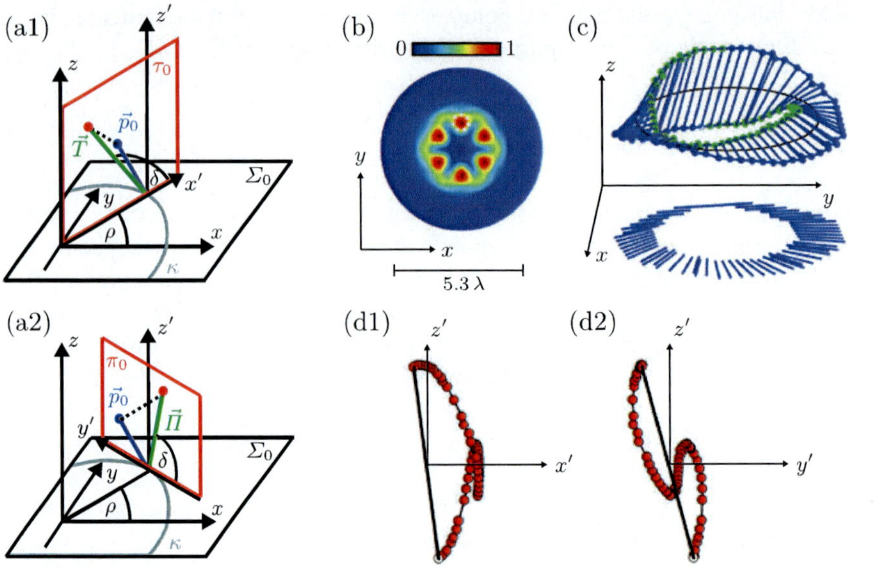

Fig. 4.13 Topological structures around off-axis generic C-points. **a** Sketch of projection planes ((a1) radial, (a2) tangential) for topology characterization. **b** Total focal intensity $|\vec{\mathcal{E}}(x, y, 0)|^2$ (normalized) of exemplarily studied, tightly focused flower configuration with $\sigma_{12} = 8$ and phase vortex $\ell = 1$. **c** Topological Möbius structure and projection on Σ_0-plane formed by tracing major axis (blue lines with green and blue endpoints) on circle (white dashed line in **b**) around top generic C-point in **b**. **d** Analysis of projections (d1) $\vec{T}$ and (d2) $\vec{\Pi}$ on radial and tangential plane, respectively (red: endpoints of projections). [Adapted with permission from Ref. [13], licensed under CC BY 4.0, https://creativecommons.org/licenses/by/4.0/]

indices $\tau_p = \pm 1/2$ or $\pm 3/2$ and $\pi_p = \pm 1/2$ or $\pm 3/2$. As major axis are π-rotation invariant, half integer values are possible for τ_p and π_p.

First, for the focused vectorial flower configuration with additional phase vortex ($\sigma_{12} = 8$, $\ell = 1$), the top C-point is selected for analysis purposes. In Fig. 4.13b, the chosen circle κ (white dashed line) around the generic point within the Σ_0- or $(x,\ y)$-plane is sketched onto the total focal intensity $|\vec{\mathcal{E}}(x, y, 0)|^2$ (normalized). Tracing the major axis $\vec{p}_0$ (blue lines with blue and green endpoints; unit length) on this circle reveals a half-twist Möbius strip, as illustrated in Fig. 4.13c. In Fig. 4.13c, the projection of major axes onto the Σ_0-plane is additionally presented. Analyzing this projection reveals a lemon-like structure which corresponds to indices $I = 1$ and $\Lambda = 1$. In order to determine indices τ_p and π_p corresponding to the radial and tangential plane, respective projections are calculated. Results are presented in Fig. 4.13d1 and d2, in which the red endpoints of projection $\vec{T}$ and $\vec{\Pi}$ (cf. Fig. 4.13a) are shown for increasing ρ in the $(x',\ z')$- and $(y',\ z')$-coordinate system. Here, both projections exhibit an index of $1/2$, i.e., $\tau_p = \pi_p = 1/2$. Hence, the found topological structure with its indices is perfectly consistent with the theoretical expectation for a generic C-point in a non-paraxial field (see above).

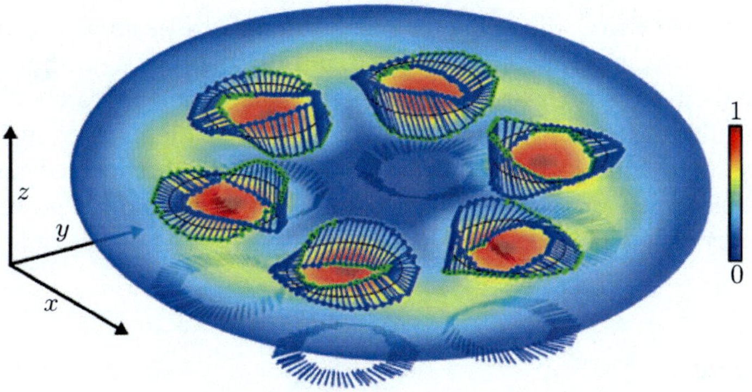

Fig. 4.14 Formation of optical polarization Möbius strip array. Shown are topological structures around all six off-axis C-points of focused tailored vector field ($\sigma_{12} = 8$, $\ell = 1$). Major axis are traced on each circle κ around singularities, presented at respective position in the total intensity structure $|\vec{\mathcal{E}}(x, y, z)|^2 \in [0, 1]$ (semitransparent). Below, projections on Σ_0-plane are given. [Adapted with permission from Ref. [13], licensed under CC BY 4.0, https://creativecommons.org/licenses/by/4.0/]

Crucially, not only an individual C-point is located within the field under consideration for $z = 0$, but a SN of six generic C-points positioned off-axis. Exploring the topology of 3d polarization states around each C-point reveals a not yet observed intriguing 3d topology network: an optical polarization Möbius strip array [13]. Respective analysis results are visualized in Fig. 4.14. Here, the traced major axes on each circle κ are illustrated at the respective position within the total focal intensity $|\vec{\mathcal{E}}|^2$ (semitransparent, $z = 0$). Further, projections onto Σ_0 are shown. Results reveal that around each C-point a half-twist Möbius strip is detected, all characterized by the same indices $I = \Lambda = 1$ and $\tau_p = \pi_0 = 1/2$. To form a continuous field of 3d polarization states, the ζ-lines of each Möbius strip point radially away from the optical axis. Since the number of Möbius strips is equal to the number of generic C-points, which can be tailored on demand by the chosen input index σ_{12}, the Möbius strip array within the non-paraxial field can be controlled by the customization of the incident vectorial field.

4.4.3 Conclusion

In total, the control of generic L- and C-singularities as well as intriguing topological structures was proven leading to the creation of not yet observed non-paraxial Möbius strip arrays [13]. By tightly focusing a tailored vectorial field of order σ_{12} and with additional phase vortex of charge ℓ, the ability to tailor the number and position of non-paraxial 3d polarization singularities was proven. Here, indices σ_{12} and ℓ serve as well-defined customization tools, being directly related to the resulting singular field properties. Beyond, the 3d polarization topology in the focal plane strongly depends on chosen input field indices: On the one hand, major axis around the optical axis may form, for example, optical cones or multi-twisted ribbons with the number of twist being given by $|\sigma_{12} - 2|$. On the other hand, topologies around off-axis C-point singularities reveal half-twist Möbius strips. As the number and position of C-points can be adjusted by the choice of σ_{12} and ℓ, the creation and customization of an extended network of Möbius strips, i.e., Möbius strip arrays, is enabled by the chosen approach to tailor focal 4d fields.

References

1. Quabis S et al (2000) Focusing light to a tighter spot. Opt Commun 179
2. Dorn R, Quabis S, Leuchs G (2003) Sharper focus for a radially polarized light beam. Phys Rev Lett 91: 233901
3. Zhan Q (2009) Cylindrical vector beams: from mathematical concepts to applications. Adv Opt Photon 1: 1
4. Otte E, Tekce K, Denz C (2017) Tailored intensity landscapes by tight focusing of singular vector beams. Opt Express 25: 20194
5. Chen W, Zhan Q (2006) Three-dimensional focus shaping with cylindrical vector beams. Opt Commun 265: 411
6. Bokor N, Davidson N (2006) Generation of a hollow dark spherical spot by 4pi focusing of a radially polarized Laguerre-Gaussian beam. Opt Lett 31: 149
7. Wang H et al (2008) Creation of a needle of longitudinally polarized light in vacuum using binary optics. Nat Photon 2: 501
8. Qin F et al (2015) Shaping a subwavelength needle with ultra-long focal length by focusing azimuthally polarized light. Sci Rep 5.9977
9. Freund I (2005) Cones, spirals, and Möbius strips, in elliptically polarized light. Opt Commun 249: 7
10. Bauer T et al (2015) Observation of optical polarization Möbius strips. Science 347: 964
11. Otte E, Tekce K, Denz C (2018) Customized focal light landscapes by complex vectorial fields for advanced optical trapping. Proc. SPIE 10549: 115
12. Otte E et al (2019) Polarization nano-tomography of tightly focused light landscapes by self-assembled monolayers. Nat Commun 10.4308
13. Tekce K, Otte E, Denz C (2019) Optical singularities and Möbius strip arrays in tailored non-paraxial light fields. Opt Express 27: 29685
14. Richards B, Wolf E (1959) Electromagnetic diffraction in optical systems. II. Structure of the image field in an aplanatic system. In: Proceedings of the royal society of London A: mathematical, physical and engineering sciences, vol 253. 1274. The Royal Society (1959), pp 358–379
15. Boruah B, Neil M (2009) Focal field computation of an arbitrarily polarized beam using fast Fourier transforms. Opt Commun 282: 4660
16. Novotny L, Hecht B (2012) Propagation and focusing of optical fields, vol 2, Chap. 3. Cambridge University Press (2012), pp 45–85
17. Novotny L, Hecht B (2012) Principles of nano-optics. Cambridge University Press (2012)
18. Gu M (2000) Advanced optical imaging theory, vol 75. Springer Science & Business Media
19. Youngworth K, Brown T (2000) Focusing of high numerical aperture cylindrical vector beams. Opt Express 7: 77
20. Biss DP, Brown TG (2001) Cylindrical vector beam focusing through a dielectric interface. Opt Express 9: 490
21. Zhan Q, Leger JR (2002) Microellipsometer with radial symmetry. Appl Opt 41: 4630

22. Pang X, Visser T, Wolf E (2011) Phase anomaly and phase singularities of the field in the focal region of high-numerical aperture systems. Opt Commun 284: 5517
23. Ashkin A (1970) Acceleration and trapping of particles by radiation pressure. Phys Rev Lett 24: 156
24. Ashkin A (2000) A History of optical trapping and manipulation of small-neutral particle, atoms, and molecules. IEEE J Sel Top Q Electron 6: 841
25. Padgett M, Bowman R (2011) Tweezers with a twist. Nat Photon 5: 343
26. Calzaferri G (2012) Nanochannels: hosts for the supramolecular organization of molecules and complexes. Langmuir 28: 6216
27. BrüChwiler D, Calzaferri G (2004) Molecular sieves as host materials for supramolecular organization. Microporous Mesoporous Mater 72: 1
28. Veiga-Gutiérrez M et al (2012) Optical-tweezers assembly-line for the construction of complex functional zeolite L structures. Adv Mater 24: 5199
29. Grundmann M (2002) Nano-optoelectronics: concepts, physics and devices. Springer Science & Business Media
30. Ashammakhi N et al (2018) Advances and future perspectives in 4D bioprinting. Biotechnol J 13: 1800148
31. Mota AH et al (2017) Broad overview of engineering of functional nanosystems for skin delivery. Int J Pharm 532. SFNano 2016 meeting
32. Rong L et al (2018) Biomedical applications of functional peptides in nano-systems. Mater Today Chem 9: 91
33. Ozin GA, Arsenault A (2015) Nanochemistry: a chemical approach to nanomaterials. Royal Society of Chemistry
34. Cademartiri L, Ozin GA (2009) Concepts of nanochemistry. Wiley
35. Mura S, Nicolas J, Couvreur P (2013) Stimuli-responsive nanocarriers for drug delivery. Nat Mater 12: 991
36. Novotny L et al (2001) Longitudinal field modes probed by single molecules. Phys Rev Lett 86: 5251
37. Rotenberg N, Kuipers L (2014) Mapping nanoscale light fields. Nat Photon 8: 919
38. Bauer T et al Nanointerferometric amplitude and phase reconstruction of tightly focused vector beams. Nat Photon 8: 23
39. Schwartz DK (2001) Mechanisms and kinetics of self-assembled monolayer formation. Ann Rev Phys Chem. 52: 107
40. Wang M et al (2005) Self-assembled silane monolayers: fabrication with nanoscale uniformity. Langmuir 21: 1848
41. Ulman A (1996) Formation and structure of self-assembled monolayers. Chem Rev 96: 1533
42. Onclin S, Ravoo BJ, Reinhoudt (2005) Engineering silicon oxide surfaces using Self-Assembled monolayers. Angew Chem Int Ed 44: 6282
43. Otte E et al (2015) Complex light fields enter a new dimension: holographic modulation of polarization in addition to amplitude and phase. Proc SPIE 9379. 937908-937908
44. Wang C, Zhang Z, Wang Y (2016) Quinacridone-based π-conjugated electronic materials. J Mater Chem C 4: 9918
45. Bauer T et al (2016) Optical polarization Möbius strips and points of purely transverse spin density. Phys Rev Lett 117: 013601
46. Berry MV (2004) Index formulae for singular lines of polarization. J Opt A: Pure Appl Opt 6: 675
47. Mermin ND (1979) The topological theory of defects in ordered media. Rev Mod Phys 51: 591
48. Dennis MR (2002) Polarization singularities in paraxial vector fields: morphology and statistics. Opt Commun 213: 201
49. Nye JF (1983) Monstars on glaciers. J Glaciol 29: 70
50. Freund I (2010) Optical Möbius strips in three-dimensional ellipse fields: I. Lines of circular polarization. Opt Commun 283: 1
51. Nye JF (1999) Natural focusing and fine structure of light: caustics and wave dislocations. CRC Press

52. Dennis MR (2001) Topological singularities in wave fields. PhD thesis. University of Bristol
53. Nye JF (1997) Line singularities in wave fields. English. Philos Trans: Math Phys Eng Sci 355: 2065
54. Freund I (2010) Optical Möbius strips in three dimensional ellipse fields: II. Lines of linear polarization. Opt Commun 283: 16

Chapter 5
Entanglement in Classical Light

Within previous chapters the ability to customize complex singular light fields in its various properties has been demonstrated. Intriguing light fields structured in 2d, 3d or even 4d have been presented giving new insights into the fundamental properties of singular light and paving the way to pathbreaking applications, e.g., in optical micromanipulation or to the implementation of functional 4d materials. However, there is another property enlarging the already rich phenomenology of structured light and opening up new perspectives for applied optics: so-called *classical entanglement*.

Entanglement is today considered as a property typically associated to quantum mechanics. This property is essential for prominent quantum observations and studies as the Gedankenexperiment of Schrödinger's cat [1, 2], the Einstein–Podolsky–Rosen paradox [3], Bell's inequality [4], quantum cryptography [5], or quantum computing [6, 7]. However, it became clear that the algebraic concept underlying entanglement can indeed be created in classical optics, facilitating the classical analogon of quantum entanglement, thus, bridging a gap between classical and quantum optics. The analogon displays essential quantum mechanical features of entanglement as the key feature of non-separability. For example structured light, namely, vectorial beams can be described as classically entangled, being non-separable in its polarization and spatial shape, as it will be shown in the following section. However, the analogy reach its limits if non-locality is considered. Hence, while local entanglement (also called local non-separability) of different degrees of freedom (DoFs, = subsystems) of a single particle (= system) is accessible in classical and quantum optics, non-local entanglement of two distant particles (= subsystems) is an exclusive quantum property [8]. Nevertheless, as it has been demonstrated within different studies [9–15], the notion of classical, thus, local entanglement may assist to infuse the potentialitites offered by quantum physics to classical optics or vice versa and, therefore, provide access to innovative sophisticated applications. For instance, classical entanglement has been exploited for real-time quantum error correction [9] and communication [16–19].

E. Otte, *Structured Singular Light Fields*, Springer Theses,
https://doi.org/10.1007/978-3-030-63715-6_5

Although beneficial features of local entanglement in classical light were proven, till now, the potential of fusing valuable quantum properties with classical ones has not been fully exploited yet. Essentially advancing this field of research, within this chapter, the ability to customize complex paraxial[1] singular fields in 3d space is studied in combination with the property of classical local entanglement. Thereby, first, the valuable properties of entanglement in classical light are outlined (Sect. 5.1). By adapting not yet considered DoFs of respective light fields, a novel means for the optimization of entanglement implementation is proposed (Sect. 5.2). For this purpose, previously introduced non-diffracting vector Bessel modes (cf. Sect. 3.3.1) are applied, facilitating the self-healing of entanglement obstructured by absorbing objects. Based on these findings [20], the combination of classical self-healing and local quantum entanglement properties allows realizing, e.g., high-dimensional quantum key distribution (QKD) through obstacles, as it has been demonstrated in Ref. [21]. Furthermore, a new light is shed onto local entanglement by presenting an approach to realize entanglement beating [22] in free space (Sect. 5.3). Within this study, a not yet known type of spin-orbit (SO) coupling, namely, paraxial SO coupling in free space, is found. The light fields embedding the entanglement oscillation are structured in 3d space, whereby these fields and their respective innovative generation technique represent advanced tools for, e.g., high-resolution imaging, optical tweezers, material processing, or QKD.

Note that the results presented in this chapter have in parts been published in Refs. [20, 22] in cooperation with the group of Prof. Dr. Andrew Forbes, University of the Witwatersrand, South Africa.

5.1 Local Entanglement in Structured Light

Generally, entanglement can be divided into two different types: the non-local entanglement of two subsystems being spatially separated from each other, and the local entanglement of the internal DoFs (= subsystems in the same spatial position) of a single system. While the former is only applicable in quantum context, the latter is also realizable in classical cases. As the property of non-locality cannot be considered for the classical cases, local entanglement is also refered to as local non-separability. A prominent example for this classical entanglement is given by spatially structured vectorial light fields in the paraxial regime, in which the polarization and spatial DoFs are non-separable, i.e., locally entangled. These examples will be elucidated in the following.

[1] In this chapter paraxial light fields of 2d polarization are considered. Thus, the term "3d (structured) fields" refers to light fields of 2d polarization, which are shaped in 3d space.

5.1.1 Non-separability of Spatial and Polarization Degrees of Freedom

Two subsystems A and B are considered as locally entangled or non-separable if the overall system $|\Psi\rangle$ cannot be represented as a factorizable product of the states $|\Psi_{A,B}\rangle$ of the subsystems, i.e., $|\Psi\rangle \neq |\Psi_A\rangle \otimes |\Psi_B\rangle$. Thereby, the symbol $\otimes$ denotes the tensor product of the two states, which is, for simplicity, omitted in the following description ($|\Psi_A\rangle \otimes |\Psi_B\rangle = |\Psi_A\rangle|\Psi_B\rangle$). For structured light, the subsystems may represent the DoFs, namely, the polarization and spatial shape (amplitude and phase) of light, being non-separable in the overall vectorial field $|\Psi\rangle$. Assuming circular polarization basis, in Dirac's notation, these kinds of classical fields are described as [23]

$$|\Psi\rangle = \sqrt{a}\,|u_R\rangle|R\rangle + \sqrt{1-a}\,|u_L\rangle|L\rangle. \tag{5.1}$$

Here, the kets $|u_{R,L}\rangle$ incorporate the spatial shape (infinite dimensional Hilbert space $\mathcal{H}_\infty$), i.e., amplitude and phase, corresponding to the right- ($|R\rangle$) or left-handed ($|L\rangle$) circular polarization basis (qubit Hilbert space $\mathcal{H}_2$). These spatial modes satisfy the normalization condition $\langle u_{L,R}|u_{R,L}\rangle = 0$ and their relative weighting is defined by a.

The best known example of locally entangled vectorial fields might be the class of cylindrical vector beams (CVBs). As outlined in Sect. 2.3.2, CVBs can be considered as a superposition of two orthogonally polarized helical Laguerre-Gaussian (LG) modes $\mathrm{LG}_{n,\ell}$ ($= \mathrm{LG}^h_{n,\ell}$, cf. Eq. (2.14)) of opposite topological charge ℓ, namely (cf. Eq. (2.26))

$$\vec{E}_{\mathrm{CVB}} = \frac{1}{\sqrt{2}}\left[\vec{e}_R \cdot \mathrm{LG}_{n,\ell} \cdot \mathrm{e}^{\mathrm{i}\alpha} + \vec{e}_L \cdot \mathrm{LG}_{n,-\ell} \cdot \mathrm{e}^{-\mathrm{i}\alpha}\right]. \tag{5.2}$$

The comparison of Eqs. (5.1) and (5.2) reveals that

$$|u_{R,L}\rangle = |\mathrm{LG}_{n,\pm\ell}\rangle\,\mathrm{e}^{\pm\mathrm{i}\alpha}, \qquad a = 1/2,$$
$$|R\rangle = \vec{e}_R = [1,\ -\mathrm{i}]^T/\sqrt{2}, \qquad |L\rangle = \vec{e}_L = [1,\ \mathrm{i}]^T/\sqrt{2}.$$

Hence, CVBs fulfill Eq. (5.1) and are, thus, non-separable or locally entangled classical states. An example for these is illustrated in Fig. 5.1a. The shown radially polarized light field represents a non-separable combination of a right-circularly polarized LG mode of charge $\ell = 1$ and a left-circularly polarized LG mode of charge $\ell = -1$. More precisely, $|u_R\rangle = |\mathrm{LG}_{0,1}\rangle$ (1. summand in Fig. 5.1a) and $|u_L\rangle = |\mathrm{LG}_{0,-1}\rangle$ (2. summand in Fig. 5.1a) with $\alpha = 0$, thus,

$$|\Psi_{\mathrm{radial}}\rangle = \frac{1}{\sqrt{2}}\left(|\mathrm{LG}_{0,1}\rangle|R\rangle + |\mathrm{LG}_{0,-1}\rangle|L\rangle\right). \tag{5.3}$$

Note that in the case of CVBs the amplitude of LG modes and, thus, the radial DoF can be fully separated from the other DoFs. More precisely, following the Eqs. (2.14)–

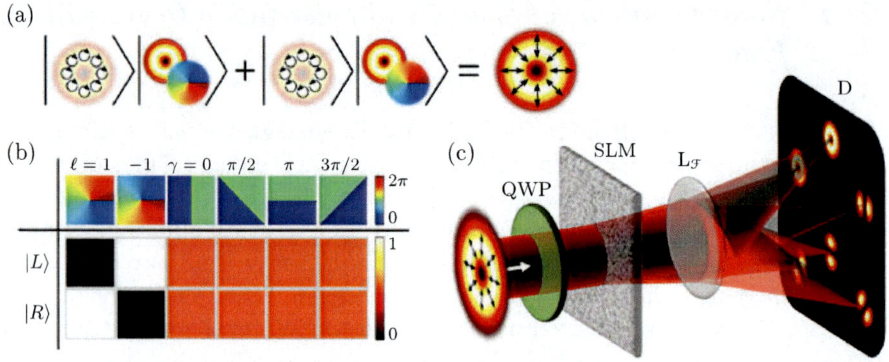

Fig. 5.1 The concept of local entanglement in classical light fields. **a** Sketch of radial CVB being non-separble in its polarization DoF and spatial shape. **b** Expected intensity values (bottom rows) resulting from projection measurements (top row: holograms for spatial mode projection) performed for a radial cylindrical vector beam (CVB) by the experimental system illustrated in (c) (QWP: quarter wave plate, SLM: phase-only spatial light modulator, $L_{\mathcal{F}}$: Fourier lens, D: detector)

(2.16) for LG beams, the right- as well as left-circularly polarized mode can be considered as

$$|u_{R,L}\rangle = |\mathrm{LG}_{n,\pm\ell}\rangle\, \mathrm{e}^{\pm \mathrm{i}\alpha} = A_{n,|\ell|} \cdot \mathrm{e}^{\mathrm{i}\zeta_{R,L}^{n,|\ell|}} \cdot |\, \mathrm{e}^{\pm \mathrm{i}\ell\varphi}\rangle. \tag{5.4}$$

Thereby, prefactors of the ket $|\exp(\pm \mathrm{i}\ell\varphi)\rangle$ are independent of the sign of the topological charge. As the amplitude $A_{n,|\ell|}$ only depends on $|\ell|$, and, thus, is equal for both circular polarization bases, it can be separated from non-separable phase (orbital angular momentum, OAM) and polarization DoF.

5.1.2 *Characterizing the Degree of Local Entanglement*

5.1.2.1 Theoretical Description

Similar to quantum mechanics [24, 25], also for classically entangled fields the degree of local entanglement or non-separability can be quantified [23], which is closely related to the parameter a in Eq. (5.1). For details on the following description see Ref. [23]. For a pure bipartite system as presented above the degree of (classical) entanglement can be defined by the von Neumann (entanglement) entropy [25]

$$E_{\mathrm{vN}} = -\mathrm{Tr}\left[\rho_p \log(\rho_p)\right], \in [0,\ 1] \tag{5.5}$$

of the reduced density matrix ρ_p of the polarization DoF [23, 26]. The density matrix is obtained by tracing over the spatial DoF (S) with

$$\rho_p = \mathrm{Tr}_S\left[\,|\Psi\rangle\langle\Psi|\,\right] \tag{5.6}$$

$$= \begin{pmatrix} a & \sqrt{a(1-a)}\,|u_L\rangle\langle u_R| \\ \sqrt{a(1-a)}\,|u_R\rangle\langle u_L| & 1-a \end{pmatrix}. \tag{5.7}$$

By this approach, a tool to measure the degree of entanglement is given with respect to the degree of mixedness of the polarization DoF of the considered state. The reduced density matrix gives the average state of polarization of the non-separable light field and is calculated from the components $s_i = \mathrm{Tr}[\varsigma_i \rho_p]$, $i = \{1, 2, 3\}$ of the Bloch vector $\vec{s}$. Note that the terms $\rho_p = (\mathbf{1} + \sum_i s_i \varsigma_i)/2$ correspond to the Stokes parameters and ς_i represent the traceless Pauli operators [23]. Calculations show that the length of the Bloch vector

$$s(\rho_p) = \left(\mathrm{Tr}\left[\rho_p^2\right]\right)^{1/2} = \left(\sum_i \langle\varsigma_i\rangle^2\right)^{1/2} \tag{5.8}$$

can be applied to determine the entanglement entropy [23] according to

$$E_{\mathrm{vN}}(|\Psi\rangle) = h\left(\frac{1+s}{2}\right), \qquad h(*) = -*\log_2(*) - (1-*)\log_2(1-*). \tag{5.9}$$

For a locally entangled classical state following Eq. (5.1), e.g., a CVB, this equation results in the expression

$$E_{\mathrm{vN}}(|\Psi\rangle) = -\left[a\log_2(a) + (1-a)\log_2(1-a)\right]. \tag{5.10}$$

Hence, if a CVB following Eq. (5.2) is considered to be radially polarized (Eq. (5.3), the degree of entanglement is given by $E_{\mathrm{vN}} = 1$ as $a = 1/2$. In contrast, a pure scalar mode is fully separable with the degree $E_{\mathrm{vN}} = 0$, as one of the summands in Eq. (5.1) disappears in this case ($a = \{0, 1\}$).

Besides the entanglement entropy, one can also rely on the closely related concurrence $C = (1 - s^2)^{1/2}$ [25, 27] as a measure for the degree of entanglement or non-separability. The real part of it defines the so-called *vector quality factor* (VQF), specifically derived for locally entangled vectorial light fields [27] with

$$\mathrm{VQF} = \Re(C) = \Re\left(\sqrt{1-s^2}\right), \in [0, 1]. \tag{5.11}$$

Also in this case, VQF = 1 corresponds to a fully entangled (non-separable) vector state,[2] whereas VQF = 0 describes a pure scalar, thus, separable state.

[2]Note that these vector states (also called vector modes/fields in the following) refer to vectorial fields of pure linear spatially varying polarization. For a maximal degree of non-separability between OAM and circular polarization, incoherently combined scalar spatial modes need to be of orthogonal shape, fulfilling the normalization condition. This restriction only allows for linearly polarized vector beams, following the superposition principle. Elliptically polarized vectorial fields (ellipse fields) of lower entanglement degree are called semi-vector states in this chapter.

Table 5.1 Normalized intensity measurements I_{uv} for the determination of expectation values $\langle \varsigma_i \rangle$ and, subsequently, the degree of entanglement (E_{vN}, VQF)

Basis states	$\ell = 1$	-1	$\gamma = 0$	$\pi/2$	π	$3\pi/2$
Left circular $\lvert L\rangle$	I_{11}	I_{12}	I_{13}	I_{14}	I_{15}	I_{16}
Right circular $\lvert R\rangle$	I_{21}	I_{22}	I_{23}	I_{24}	I_{25}	I_{26}

5.1.2.2 Experimental Determination

In order to experimentally determine the degree of entanglement characterized as the entanglement entropy E_{vN} or vector quality factor VQF a system as illustrated in Fig. 5.1c is applied. This system consists of a quarter wave plate (QWP), a polarization sensitive spatial light modulator (SLM), a Fourier lens ($\mathrm{L}_{\mathcal{F}}$), and a detector. The light field is analyzed in the plane of the SLM, which is Fourier transformed onto the detector, e.g., a camera. This system is used to measure the expectation values of the Pauli operators $\langle \varsigma_i \rangle$ and, thus, the length of the Bloch vector s. In total, one needs to perform 12 normalized on-axis intensity measurements or six identical measurements for two different basis states to determine E_{vN} or VQF following Eq. (5.10) or (5.11), respectively [23, 27].

For circular polarization basis $|L\rangle$ and $|R\rangle$, spatial projection measurements are given by two OAM modes of charge ℓ and $-\ell$, i.e., spatial modes $|u_{R,L}\rangle = |u_{\pm\ell}\rangle$, as well as four superposition states represented by $|u_\ell\rangle + \exp(\mathrm{i}\gamma)|u_{-\ell}\rangle$ with $\gamma = \{0,\ \pi\ell/2,\ \pi\ell,\ 3\pi\ell/2\}$. According to Table 5.1, the expectation values can be calculated from

$$
\begin{aligned}
\langle \varsigma_1 \rangle &= I_{13} + I_{23} - (I_{15} + I_{25}),\\
\langle \varsigma_2 \rangle &= I_{14} + I_{24} - (I_{16} + I_{26}),\\
\langle \varsigma_3 \rangle &= I_{11} + I_{21} - (I_{12} + I_{22}).
\end{aligned}
$$

Respective on-axis intensity values I_{uv}, $u = \{1,\ 2\}$, $v = \{1,\ 2,\ \ldots,\ 6\}$, are normalized by $I_{11} + I_{12} + I_{21} + I_{22}$.

For polarization projections, the QWP is applied with its fast axis set to $\pm\pi/4$ ($\pm 45°$) and combined with the polarization filter properties of the SLM which only modulates horizontally polarized light. To perform spatial projection measurements, holograms of OAM modes $|u_{\pm\ell}\rangle$ and superposition states are displayed on the SLM.[3]

[3] A blazed grating is added to respective holograms to assign the horizontally polarized decoded light to a specific position in Fourier space, ensuring polarization filtering.

To accelerate the measurement procedure, phase-only holograms can be multiplexed as indicated in Fig. 5.1c. For this purpose, each mode is assigned to another spatial carrier frequency, i.e., blazed grating of chosen characteristics. Hence, the respective on-axis intensities are detected at different specific positions in space by, e.g., a camera.

Note that, for CVBs, spatial modes are $|u_{\pm\ell}\rangle = |\mathrm{LG}_{n,\pm\ell}\rangle \cdot \exp(\pm\mathrm{i}\alpha)$ including amplitude structuring. However, as the radial DoF, i.e., the amplitude of LG modes, is fully separable within the system, it is neglected and pure phase holograms are applied, namely $|\exp(\pm\mathrm{i}\ell\phi)\rangle$ (cf. Eq. (5.4)). As an illustrative example, the expected intensity values for a radially polarized CVB are depicted in Fig. 5.1b, arranged according to Table 5.1. In addition, the respective phase-only holograms for the OAM and superposition modes are presented in the top row with $\ell = 1$, as a CVB of lowest order is considered. Respective intensity values result in an entanglement entropy of $E_{\mathrm{vN}} = 1$ and VQF $= 1$.

Crucially, the presented approach of non-separable polarization DoF and spatial shape is not only applicable in classical optics, but also in quantum cases considering the local entanglement of the internal DoFs of a single photon. The benefit of the respective non-separability has already been proven impressively by the realization of, e.g., higher-dimensional QKD for which a set of mutually unbiased bases (MUB) is applied, formed by hybrid scalar and locally entangled vector modes on LG basis [28].

5.2 Recovery of Local Entanglement

Above, the concept of classical entanglement or non-separability was introduced by the example of CVBs. Crucially, not only LG modes can be chosen as basis $|u_{R,L}\rangle$ for classically entangled fields, but also various light fields carrying well-defined OAM of charge ℓ. This characteristic paves the way to combine the customizable properties of structured singular light fields and local non-separability—applicable in classical and quantum cases. Here, an approach is proposed by which the valuable self-healing characteristics of non-diffracting vector Bessel modes, i.e., 3d structured vectorial fields, are fused with the property of local entanglement, facilitating the recovery of the degree of entanglement upon propagation [20]. The fusion of classical and originally quantum properties is demonstrated by measuring the VQF at different levels of self-healing of non-separable vector Bessel fields. It is experimentally shown that non-separability recovers behind applied obstructions to the original degree of entanglement. Results are confirmed by additional Bell-like inequality measurements [4] in its most commonly used form for optics, i.e., Clauser-Horne-Shimony-Holt (CHSH) inequality [29]. Overall findings enable the implementation of self-healing characteristics for advanced applications as QKD through obstacles [21].

5.2.1 *Locally Entangled Vector Bessel Modes*

5.2.1.1 Shaping the Separable Radial DoF

To join the beneficial properties of Bessel modes and local non-separability, spatial modes $|u_{R,L}\rangle$ in Eq. (5.1) are replaced by self-healing transverse eigenstates, namely scalar higher-order Bessel or, experimentally, Bessel-Gaussian (BG) modes. BG modes $E_\ell^{\mathrm{BG}}(r, \varphi, z)$ were defined in Sect. 3.3.1, Eq. (3.10), with its topological charge ℓ and radial wave number k_r. As shown within this section, the incoherent superposition of two BG modes may result in self-healing vectorial BG (vBG) beams, classically described as (cf. Eq. (3.13))

$$\vec{E}^{\mathrm{vBG}} = \frac{1}{\sqrt{2}}\left[\vec{e}_R \cdot E_\ell^{\mathrm{BG}} \cdot \mathrm{e}^{\mathrm{i}\alpha} + \vec{e}_L \cdot E_{-\ell}^{\mathrm{BG}} \cdot \mathrm{e}^{-\mathrm{i}\alpha}\right]. \tag{5.12}$$

Obviously, with $|u_{k_r,\pm\ell}\rangle = E_{\pm\ell}^{\mathrm{BG}}$ this 3d structured light field can be written as

$$|\Psi_{k_r,\ell}\rangle = \frac{1}{\sqrt{2}}\left(|R\rangle|u_{k_r,\ell}\rangle \cdot \mathrm{e}^{\mathrm{i}\alpha} + |L\rangle|u_{k_r,-\ell}\rangle \cdot \mathrm{e}^{-\mathrm{i}\alpha}\right). \tag{5.13}$$

Hence, the self-healing vBG field is identified as a non-separable state of light with $|u_{R,L}\rangle = |u_{k_r,\pm\ell}\rangle \exp(\pm\mathrm{i}\alpha)$ and $E_{\mathrm{vN}} = \mathrm{VQF} = 1$. Crucially, the self-healing property is traditionally attributed to the radial component of a spatial mode, which is, also for vBG modes, fully separable, namely

$$|\Psi_{k_r,\ell}\rangle = \frac{1}{\sqrt{2}} E_{k_r,|\ell|}^{\mathrm{BG}}(r)\left(|R\rangle|\ell\rangle \cdot \mathrm{e}^{\mathrm{i}\alpha} + |L\rangle|-\ell\rangle \cdot \mathrm{e}^{-\mathrm{i}\alpha}\right). \tag{5.14}$$

Here, $E_{k_r,|\ell|}^{\mathrm{BG}}(r)$ is the separable radial profile and $|\pm\ell\rangle$ includes the non-separable phase information of the BG fields. Hence, by replacing the spatial modes $|u_{R,L}\rangle$ by BG fields and, thereby, shaping the separable radial profile of the total field, a non-diffracting and self-healing locally entangled light field is realized.

5.2.1.2 Experimental Realization of Obstructed Non-separable Fields

In the following, the effect of the originally classical self-healing property, encoded within the separable radial DoF, on the degree of entanglement of the overall field is studied. To experimentally realize non-separable vBG modes, the technique introduced in Sect. 3.3.1 and illustrated in Fig. 3.9c is applied. Thus, first a SLM (SLM_1) forms a fundamental BG mode ($\ell = 0$, $k_r = 18\,\mathrm{rad\,mm}^{-1}$) by a binary Bessel function (cf. Eq. (3.11)) as phase-only hologram in combination with a blazed grating and a Gaussian envelope ($w_0 = 0.89\,\mathrm{mm}$). Subsequently, a q- and wave plate(s) are applied to transform the scalar to a vectorial mode. By this approach, a vBG field,

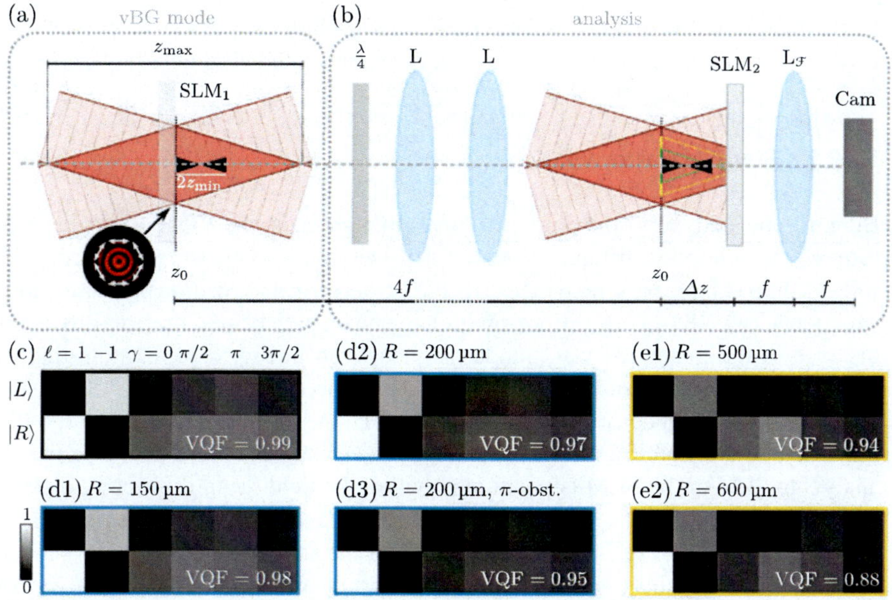

Fig. 5.2 Investigation of non-separability in self-healing vBG beams. **a**, **b** Concept of the entanglement analysis of vBG modes obstructured by digitally realized objects of radius R (SLM: spatial light modulator, $\frac{\lambda}{4}$: quarter wave plate, $L_{(\mathcal{F})}$: (Fourier) lens of focal length f, Cam: camera). **c–e** Measured on-axis normalized intensity values I_{uv} arranged according to Table 5.1 for the **c** non-obstructed as well as **d**/**e** obstructed case. Obstacles **d** allow for (blue)/**e** impede (yellow) full self-reconstruction of vBG mode before being analyzed. Determined VQFs are shown within subfigures. [Adapted with permission from Ref. [20]]

as depicted in Fig. 5.2a, is realized with its z_0-plane being at the image plane of the generating SLM$_1$. As demonstrated in Sect. 3.3.1, besides generating the field, the SLM is simultaneously able to create absorbing or phase obstacles, being adaptable all-digitally and located in the z_0-plane of the vBG mode. The following study focuses on on-axis obstacles of radius R.

5.2.2 Self-healing Degree of Local Entanglement

To study the combination of self-healing properties and local entanglement in more detail, a representative vBG field is obstructured by tailored obstacles and its non-separability is analyzed for different levels of self-healing. For this purpose, the plane of analysis, i.e., the detection plane is chosen to be fixed and located $\Delta z = 23$ cm from the z_0-plane. As sketched in Fig. 5.2b (detection plane = SLM$_2$), by this choice different levels of self-healing are accessible by simply changing the size R of the obstacle [20]. For smaller obstacles (black shadow/reconstruction region) for which the self-healing distance $z_{\min}$ fulfills $2z_{\min} \leq \Delta z$, a fully-reconstructed field is analyzed. In contrast, for larger obstacles (yellow or green dashed lines) with $2z_{\min} > \Delta z$ not fully recovered, i.e., partially reconstructed vBG fields are examined.

5.2.2.1 Study of Vectorness

Here, for the analysis of the degree of local entanglement, the VQF, also called vectorness, is studied for different levels of self-healing. The measurement procedure follows the principle presented in Sect. 5.1.2. To perform polarization projection measurements, a QWP ($\frac{\lambda}{4}$) is positioned within the beam path behind the non-diffracting distance of the generated vBG mode, as visualized in Fig. 5.2a, b. Note that the position of the QWP can be chosen as desired, since a paraxial light field is studied. By a $4f$-system constituted of two lenses (L) the (a) realized vBG field, which may include a tailored digital obstruction, is (b) imaged onto a plane before the analyzing SLM_2. The SLM is positioned within the field's non-diffracting distance $z_{\max} = 49.16$ cm (for wavelength $\lambda = 633$ nm) but Δz behind the image plane, making different levels of self-healing accessible for the analysis. SLM_2 displays the six multiplexed holograms used for spatial projection measurements, namely, two OAM modes and four superposition states. Crucially, here, spatial modes $|u_{R,L}\rangle = |u_{\pm\ell}\rangle$ constituting decoding holograms represent scalar BG modes $E_\ell^{\mathrm{BG}}(r, \varphi, z)$. However, as the radial component is entirely separable, it is omitted in the holograms such that $|\ell\rangle, |-\ell\rangle$, and $|\ell\rangle + \exp(\mathrm{i}\gamma)|-\ell\rangle$ with $\gamma = \{0, \pi\ell/2, \pi\ell, 3\pi\ell/2\}$ are applied (cf. Eq. (5.14)). Experimentally, $|\pm\ell\rangle$ is represented by the phase-only binary Bessel function hologram. Optically Fourier transforming the observation plane (= SLM_2) onto a camera enables the detection of on-axis intensity values I_{uv}. As holograms are multiplexed, for each polarization projection, the intensity values are measured by a single shot, enabling the determination of expectation values $\langle \varsigma_i \rangle$ for the calculation of VQF. Following this approach, by including obstacles of different radius R through SLM_1, the VQF can be determined at different levels of self-healing.

As representative example, an azimuthally polarized vBG mode is formed, facilitated by a q-plate of charge $q = 1/2$ (cf. Sect. 3.3.1). Hence, incorporated scalar BG fields are of topological charge $\ell = \pm 1$, so that projections are accordingly performed for $\ell = \pm 1$. Following Eq. (5.13), the respective light field is described by

$$|\Psi_{k_r,1}\rangle = \frac{1}{\sqrt{2}}\left(|R\rangle|u_{k_r,1}\rangle - |L\rangle|u_{k_r,-1}\rangle\right)$$

($\alpha = \pi/2$). First, the maximal non-separability for this azimuthally polarized vBG mode is demonstrated without obstacle by measuring the VQF as described above. Arranged according to Table 5.1, measured normalized intensity values I_{uv} are demonstrated in Fig. 5.2c. Based on these results, a VQF of 0.99 is determined for the studied vBG mode, proving the expectation for this non-separable light field. Investigating the interplay of self-healing and local entanglement, in a next step, differently sized obstacles are imparted digitally and the VQF is determined. On the one hand, obstacles of radius $R = \{150, 200\}$ µm are embedded within the z_0-

plane of the vBG field. Here, two absorbing ($R = \{150, 200\}\,\mu$m) and one phase obstacle (π-obstacle, $R = 200\,\mu$m), causing a homogeneous phase shift of π, are considered. For these obstacles the light field is fully reconstructed before being decomposed by SLM_2, i.e., $2z_{\min} < \Delta z$ with $2z_{\min} = 16.54$ cm for $R = 150\,\mu$m and $2z_{\min} = 22.06$ cm for $R = 200\,\mu$m. On the other hand, absorbing obstacles of radius $R = \{500, 600\}\,\mu$m are programmed, impeding the full reconstruction of the vBG mode before SLM_2. In these cases, $2z_{\min} > \Delta z$ (even $z_{\min} > \Delta z$) with $2z_{\min} = 55.14$ cm for $R = 500\,\mu$m and $2z_{\min} = 66.06$ cm for $R = 600\,\mu$m. Measurement results are presented in Fig. 5.2d, e for the fully reconstructed fields (blue) and partially self-healed cases (yellow), respectively. Determined VQFs are given within each subfigure.

Results reveal that the degree of local entanglement decreases with increasing obstacles size, thus, with decreasing level of self-healing [20]. If the beam is obstructed by absorbing obstacles but can fully reconstruct before its decoding (Fig. 5.2d1/d2), the VQF only differs minimally from the maximal degree of entanglement measured for the non-obstructed vBG mode (Fig. 5.2c). For the phase obstruction, the VQF is slightly lower, since, in contrast to absorbing obstacles causing a loss of information, the phase obstacle do not cut but vary information. The loss of information for absorbing obstacles can be compensated within $2z_{\min}$ by plane waves passing the obstacle. As opposed to this, varying information by a phase obstacle adds undesired information, i.e., noise to the vBG field. This noise is not eliminated upon propagation, thus, the beam stays disturbed also at the detection plane, so that the measured VQF decreases.

Compared to the self-healed cases, if the vBG mode is not able to fully reconstruct before being analyzed Fig. 5.2e, a relativly large decrease in VQF is observed [20]. However, for fully as well as partially reconstructed vectorial modes the degree of non-separability is VQF ≥ 0.88. Hence, in all cases, the light field is much closer to being vector or non-separable (VQF $= 1$) than scalar or separable (VQF $= 0$). This effect is explained by the fact that there is always undisturbed information carried by obstacle passing plane waves reaching the decoding SLM_2. As conclusion, the vBG fields can always be considered as non-separable, i.e., locally entangled, whereby obstacle based noise causes a decrease of the measured VQF quantifying the non-separability. Since for absorbing obstacles the noise is annihilated upon propagation, this quantitative value recovers. The found phenomenon is similar to what is known as "self-healing" of amplitude, phase and polarization of non-diffracting vectorial fields: obstructions add noise to the DoFs or cause a loss of information. However, with propagation distance, the respective perturbation vanishes so that the undisturbed field information is left. Thus, the DoFs seem to self-reconstruct.

To summarize, not only amplitude, phase and polarization of an obstructed vBG mode recovers upon propagation, but also the measured degree of non-separability quantified by the VQF. The larger the obstacle, i.e., the lower the level of self-healing, the smaller this quantity.

5.2.2.2 Bell-Like Inequality Measurement

As a tool for the confirmation of previous finding on the self-healing properties of non-separability or, more precisely, the VQF [20], the Bell parameter is applied [23, 30]. As the most commonly used Bell-like inequality for optical systems, Clause-Horne-Shimony-Holt (CHSH) inequality measurement is performed for the quantification of local entanglement between polarization and spatial DoFs. Instead of the non-local measurement of a single DoF (polarization or OAM), two DoFs (polarization and OAM) are studied locally within the same classical light field. As outlined in Ref. [23], this procedure is valid for the case at hand due to the analogy of quantum and classical local entanglement. To perform the respective measurement, the QWP in the experimental analysis system (Fig. 5.2b) is replaced by a half wave plate (HWP, $\frac{\lambda}{2}$). Still, the plane of analysis is fixed at Δz behind the z_0-plane, where SLM_2 is positioned. To quantify the non-separability of the exemplary azimuthally polarized (obstructed) vBG mode, the on-axis intensity $I(\theta_A, \theta_B)$ is recorded for different angles $2\theta_A = \{0, \pi/8, \pi/4, 3\pi/8\}$ of the HWP and rotation angles $\theta_B \in [0, \pi]$ of the hologram encoded on SLM_2. The hologram includes the information of spatial modes embedded in the vBG field, more precisely, it is defined as $|u_{k_r,\ell}\rangle + \exp(\mathrm{i}2\theta_B)|u_{k_r,-\ell}\rangle$ ($\ell = 1$, $k_r = 18\,\mathrm{rad\,mm}^{-1}$). Also for this measurement, due to the separability of the radial component in vBG field, spatial modes $|u_{k_r,\pm\ell}\rangle$ are substituted by $|\pm\ell\rangle$. Experimentally, $|\pm\ell\rangle$ is realized as phase-only hologram based on the binary Bessel function. The respective measurement enables the calculation of the CHSH-Bell parameter S according to [23]

$$S = E(\theta_A, \theta_B) - E(\theta_A, \theta'_B) + E(\theta'_A, \theta_B) + E(\theta'_A, \theta'_B), \tag{5.15}$$

with $E(\theta_A, \theta_B)$ being calculated from measured on-axis intensity values with

$$\begin{aligned} E(\theta_A, \theta_B) &= \frac{A(\theta_A, \theta_B) - B(\theta_A, \theta_B)}{A(\theta_A, \theta_B) + B(\theta_A, \theta_B)}, \\ A(\theta_A, \theta_B) &= I(\theta_A, \theta_B) + I\left(\theta_A + \frac{\pi}{2}, \theta_B + \frac{\pi}{2}\right), \\ B(\theta_A, \theta_B) &= I\left(\theta_A + \frac{\pi}{2}, \theta_B\right) + I\left(\theta_A, \theta_B + \frac{\pi}{2}\right). \end{aligned} \tag{5.16}$$

For separable states $|S| \leq 2$, whereas for entangled or non-separable states the maximum value for the Bell paramter is given by $|S| = 2\sqrt{2}$, known as the Tsirelson's bound [31]. In Fig. 5.3 results are depicted with respective intensity points and fitting curves presented in graphs per incorporated obstacle. Respective curves are applied for the calculation of $|S|$ following Eqs. (5.15) and (5.16). The values are presented within each subfigure. Similar to the vectorness study, the graphs are determined for the exemplary azimuthal vBG mode (a) without and (b)/(c) with obstructions of different radii. Digitally embedded obstacles are of radius (b) $R = \{150, 200\}\,\mu\mathrm{m}$ and (c) $R = \{500, 600\}\,\mu\mathrm{m}$, thus, allow for (blue boxes) and impede (yellow boxes) full self-reconstruction of the vectorial light field, respectively, before being analyzed at

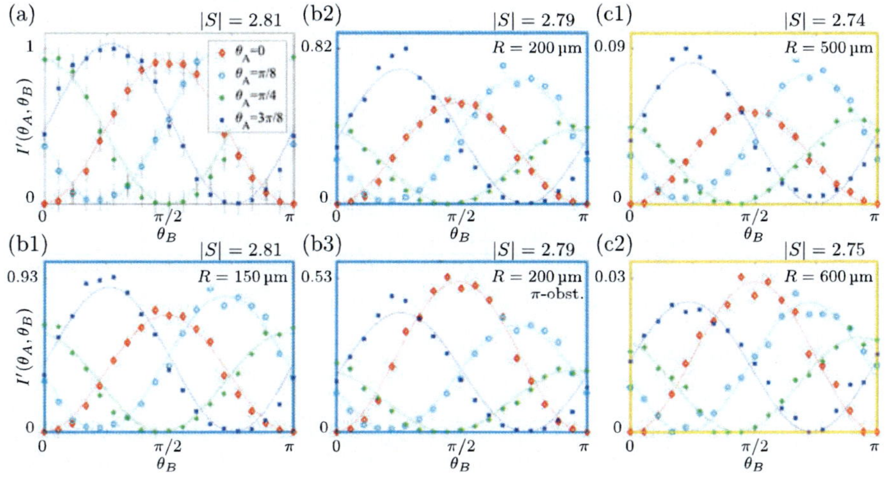

Fig. 5.3 Bell-like inequality measurement of self-healing vBG mode. Bell-type curves used to determine the Bell parameter $|S|$ are shown for an azimuthal vBG field with four different HWP orientations $2\theta_A$ and the rotation angle θ_B of the applied hologram. Measurements are performed for **a** the undisturbed case (with exemplary error bars) as well as **b/c** for differently sized obstacles of radius R allowing for/impeding (blue/yellow boxes) full self-recovery before the light field analysis **b1, b2**, **c** absorbing obstacles; **b3** π-obstacle). Measured on-axis intensity values are normalized according to the maximum intensity detected without obstacle (**a**). Bell parameters are determined by $\cos^2$-fits (dashed curves). [Adapted with permission from Ref. [20]]

SLM_2. All obstructions are absorbing ones, except for results in Fig. 5.3b3, for which the π-obstacle was applied. Note that all intensity values are normalized according to the maximum intensity detected for the field without obstruction ($I'(\theta_A, \theta_B)$: normalized values). Based on systematic and technical errors as well as the standard deviation within the experiment, an error of $\pm 7\%$ of the maximum intensity $I'_{\text{max}}(R)$ in each graph is considered, as exemplarily illustrated by error bars in Fig. 5.3a.

As the $I'(\theta_A, \theta_B)$ axes reveal, the maximally measurable intensity $I'_{\text{max}}(R)$ decreases dramatically with increasing obstacle size. Despite this effect, the respective Bell parameter $|S|$ does not change significantly depending on obstacle radius R, especially for the case of fully self-healed beams (b). In accordance with the VQF analysis, if the light field did not fully recover before its analysis, relatively bigger changes are observed for $|S|$. However, although the intensity lowers to only some percentage (c2) of the original maximum value (a), all measurement validate a violation of the Bell inequality and, therefore, confirm the previous findings on non-separability in self-healing vBG field.

Table 5.2 Quantification of local entanglement or non-separability properties of recovering vBG modes as function of the self-healing level, given by the obstruction radius R

R in μm	0	150	200	200, π-obst.	500	600
I'_{max}	1	0.93	0.82	0.53	0.09	0.03
VQF	0.99	0.98	0.97	0.95	0.94	0.88
$\|S\|$	2.81	2.81	2.79	2.79	2.74	2.75

5.2.3 Conclusion

By adapting the spatial mode being locally entangled with the state of polarization, a method was proposed for the fusion of valuable classical and originally quantum properties. Here, by the implementation of Bessel modes or, more precisely, BG fields, the property of non-diffraction and self-healing was joint with non-separability. While the non-separability is found in the azimuthal DoFs, namely, the azimuthal beam profile and polarization, the radial component is fully separable.

The self-healing of non-separability was found by studying the degree of local entanglement by the VQF as well as a classical version of the CHSH Bell-like inequality. Results are summarized in Table 5.2. Non-separable vBG modes were generated according to Sect. 3.3.1, whereby its analysis was performed via a specifically designed detection system. With this system, the variation of obstruction size was equal to changing the level of self-healing, thus, it enabled the investigation of the relation between self-healing level and non-separability [20]. Due to the Gaussian envelope of BG modes, the measurable maximum intensity $I'_{max}(R)$ shows an approximately Gaussian decrease with increasing obstacle size. Both, vector quality and Bell analysis reveal generally high degrees of non-separability (VQF and $|S|$) for all obstacle sizes. However, slight changes within VQF and $|S|$ prove the self-reconstruction of the degree of non-separability [20]. Physically, this effect can be described by the annihilation of obstacle based perturbation with the level of self-healing, i.e., propagation distance within a non-separable field. Hence, as within the non-diffracting beam lost information is compensated and noise is canceled by obstacle passing waves, quantitative values for local entanglement recover.

Crucially, the property of self-healing is typically attributed to the fully separble radial DoF. Nevertheless, as it was proven, the self-recovering characteristic is inherited to the locally entangled vBG field and its degree of non-separability. As a consequence, presented finding give an insight into the ability to tailor the unused radial DoF in order to assign valuable classical properties to locally non-separable light fields. For instance, by BG bases, local entanglement can be made more resilient to decay from obstructions—the benefit of this approach has already been demonstrated by performing QKD through obstacles with single locally entangled photons [21].

5.3 Local Entanglement Beating in Free Space

Under local unitary transformations, the entanglement of a quantum state is invariant [32]. This well-known property is true for both, non-local as well as local entanglement. For local entanglement this rule dictates that the non-separability of the internal DoFs of a single photon does not change, e.g., upon free space propagation, which is a beneficial property for applications as QKD. Due to the analogy between classical and quantum entanglement in its local form, this rule also applies for non-separable classical fields as CVBs.

In the following, a scenario is outlined in which this paradigm does not hold, giving new insight into the properties of local classical as well as quantum entanglement. Sculpting a 3d structured vectorial light field on the basis of CVBs, local entanglement is formed oscillating from maximally entangled (purely vector) to fully separable (purely scalar) upon propagation [22]. The proposed approach is based on the superposition of two non-separable CVBs counter-propagating on-axis. The entanglement beating observed within this superposition is created by a novel form of spin-orbit (SO) coupling, namely paraxial SO coupling in free space [22]. Besides providing theoretical evidence of this effect, in this section, the SO coupling and paradigm changing entanglement beating is demonstrated experimentally. For this purpose, digital counter-propagation of vector modes and the simultaneous quantification of the degree of non-separability by the entanglement entropy E_{vN} are applied. Further, respective findings offer a tool for the transport of entanglement reminiscent of tractor beams for optically trapped particles [33–36]. As outlined later on, this phenomenon is of interest for quantum and classical communication, improved material processing and switchable imaging in stimulated emission depletion (STED) microscopy.

5.3.1 *Spatially Varying Non-separability by Spin-Orbit Coupling*

5.3.1.1 Theoretical Concept

As presented in Sect. 5.1 a CVB, constituted of two orthogonally polarized LG beams of opposite topological charge, has a propagation invariant maximal degree of entanglement. This is easily seen from the respective equation, namely,

$$|\Psi_{\mathrm{CVB}}^{\pm}\rangle = \frac{1}{\sqrt{2}}\left(\mathrm{e}^{\mathrm{i}\alpha}\,|\mathrm{LG}_{n,\ell}\rangle|R\rangle + \mathrm{e}^{-\mathrm{i}\alpha}\,|\mathrm{LG}_{n,-\ell}\rangle|L\rangle\right)\cdot \mathrm{e}^{\pm\mathrm{i}k_z z} \tag{5.17}$$

in which $\exp(\pm\mathrm{i}k_z z)$ (k_z: longitudinal wave number) approximates the light field propagation in $\pm z$-direction (paraxial, approximately collimated field). Summarizing $\zeta_{R,L} = \pm\alpha \pm k_z z$, the spatial DoF of the non-separable state (cf. Eq. (5.1)) is repre-

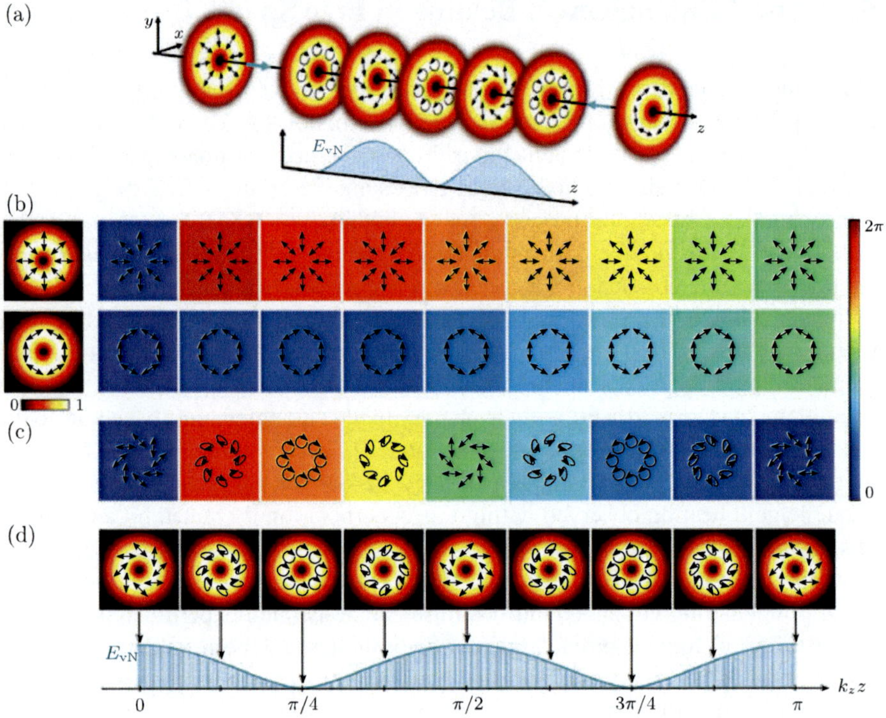

Fig. 5.4 Schematics of the studied 3d structured field embedding z-dependent entanglement beating ($\ell = 1$, $n = 0$). **a** Concept, **b** the change of phase of the two counter-propagating CVBs (top: radial CVB_1, bottom: azimuthal CVB_2) relative to the initial phase with constant polarization structure (black arrows), **c** absolute relative phase difference between CVB_1 and CVB_2 resulting in a 3d structured polarization, **d** change of polarization upon normalized intensity of $|\Psi_{3d}\rangle$ with corresponding degree of entanglement E_{vN} (blue curve), all dependent on propagation distance z or $k_z z \in [0, \pi]$ ($\delta = 0$). [Adapted with permission from Ref. [22], licensed under CC BY 4.0, https://creativecommons.org/licenses/by/4.0/]

sented by $|u_{R,L}\rangle = |\mathrm{LG}_{n,\pm\ell}\rangle \exp(\mathrm{i}\zeta_{R,L})$ with $\langle u_{L,R}|u_{R,L}\rangle = 0$. Hence, the weighting factors are defined by $a = 1/2$ so that for all propagation distances z the degree of entanglement is given by $E_{\mathrm{vN}}(|\Psi^{\pm}_{\mathrm{CVB}}\rangle) = 1$.

Remarkably, a spatially varying degree of local entanglement can be realized in a vectorial 3d light field $|\Psi_{3\mathrm{d}}(x, y, z)\rangle$ which changes its polarization structure longitudinally (in z-direction), as sketched in Fig. 5.4a. For this purpose, counter-propagating CVBs of orthogonal polarization are considered with one vector beam $|\Psi^{+}_{\mathrm{CVB}_1}\rangle$ ($\alpha_{\mathrm{CVB}_1} = 0$) propagating in $+z$- and the second one $|\Psi^{-}_{\mathrm{CVB}_2}\rangle$ ($\alpha_{\mathrm{CVB}_1} = \pi/2$) in $-z$-direction [22], i.e.,

$$|\Psi^{+}_{\mathrm{CVB}_1}\rangle = \frac{1}{\sqrt{2}}\left(|\mathrm{LG}_{n,\ell}\rangle|R\rangle + |\mathrm{LG}_{n,-\ell}\rangle|L\rangle\right)\cdot \mathrm{e}^{\mathrm{i}k_z z}, \tag{5.18}$$

$$|\Psi^{-}_{\mathrm{CVB}_2}\rangle = \frac{1}{\sqrt{2}}\left(\mathrm{e}^{\mathrm{i}\pi/2}\,|\mathrm{LG}_{n,\ell}\rangle|R\rangle + \mathrm{e}^{-\mathrm{i}\pi/2}\,|\mathrm{LG}_{n,-\ell}\rangle|L\rangle\right)\cdot \mathrm{e}^{-\mathrm{i}k_z z}. \tag{5.19}$$

Considered in $+z$-direction, i.e., overall propagation direction, these two combined CVBs reveal different changes in global phase, as exemplarily depicted in Fig. 5.4b for $\ell = 1$ and $n = 0$. Crucially, this difference in phase structures cause a z-dependent oscillation in the relative phase difference between CVB_1 and CVB_2, seen in Fig. 5.4c. As consequence, the transverse polarization configuration (black arrows) changes with propagation distance. The respective total 3d structured field is given by [22]

$$\begin{aligned} |\Psi_{3\mathrm{d}}\rangle &= \frac{1}{\sqrt{2}}\left(|\Psi^{+}_{\mathrm{CVB}_1}\rangle + |\Psi^{-}_{\mathrm{CVB}_2}\rangle\right) \\ &= \frac{1}{2}\left(\mathrm{e}^{\mathrm{i}k_z z} + \mathrm{i}\,\mathrm{e}^{-\mathrm{i}k_z z}\right)|\mathrm{LG}_{n,\ell}\rangle|R\rangle + \frac{1}{2}\left(\mathrm{e}^{\mathrm{i}k_z z} - \mathrm{i}\,\mathrm{e}^{-\mathrm{i}k_z z}\right)|\mathrm{LG}_{n,-\ell}\rangle|L\rangle. \end{aligned} \tag{5.20}$$

As visualized in (d), while the amplitude/intensity (background, normalized) of this field stays unchanged upon propagation, the polarization (black arrows) changes in such a way that the degree of entanglement E_{vN} (blue curve) varies as a function of z-distance. Following Eq. (5.10), the corresponding entanglement entropy is calculated from [22]

$$\begin{aligned} E_{\mathrm{vN}}(|\Psi_{3\mathrm{d}}\rangle, z) = 1 &- \frac{1}{2}[1 + \sin(2k_z z)]\cdot\log[1 + \sin(2k_z z)] \\ &- \frac{1}{2}[1 - \sin(2k_z z)]\cdot\log[1 - \sin(2k_z z)]. \end{aligned} \tag{5.21}$$

Hence, upon free space propagation, the engineered state undergoes a periodic oscillation between maximally non-separable ($E_{\mathrm{vN}} = 1$), i.e., vector, and fully separable ($E_{\mathrm{vN}} = 0$), i.e., scalar. Maximal non-separability is found at $z = j\lambda/4$, $j \in \mathbb{N}$, whereas complete separability or non-entanglement ($E_{\mathrm{vN}} = 0$) is observed at $z = (2j + 1)\lambda/8$, $j \in \mathbb{N}$.

The sophisticated entanglement beating provides a tool to enable the transport of a chosen degree of non-separability, hence, chosen state of light, to a predefined position in space [22]. By simply adjusting a phase factor δ the state can be shifted across arbitrary distances, reminiscent of tractor beams [33–36]. For this purpose, an additional phase factor is applied on both counter-propagating CVBs, i.e., the propagation factor $\exp(\pm\mathrm{i}k_z z)$ in Eqs. (5.18) and (5.19) is replaced by $\exp(\pm\mathrm{i}[k_z z + \delta])$. In this manner, one can move, for instance, the plane of maximal degree of entanglement ($E^{\max}_{\mathrm{vN}}(|\Psi_{3\mathrm{d}}\rangle) = 1$) [22] to a chosen position $z_{\max}$ following

$$z_{\max}(\delta) = \frac{\lambda}{4}\left(j - \frac{2\delta}{\pi}\right), \quad j \in \mathbb{Z} \tag{5.22}$$

by adjustment of δ. Beyond, by imparting a time varying phase shift $\delta(t)$, one can even realize a time dependent movement of the considered plane of $|\Psi_{3d}\rangle$. The axial velocity of this plane and its respective degree of entanglement is given by [22]

$$v_{\max}(t) = -\frac{\lambda}{2\pi}\frac{\partial\delta(t)}{\partial t}. \tag{5.23}$$

5.3.1.2 Counter-Propagation of Higher-Order CVBs and Scalar Fields

Note that the found entanglement beating cannot only be realized for $\ell = 1$, but can be shaped by any OAM subspace ℓ by superposition of orthogonally polarized CVBs $|\Psi^{+}_{\mathrm{CVB}_1}\rangle$ and $|\Psi^{-}_{\mathrm{CVB}_2}\rangle$ with radial mode number $n_{\mathrm{CVB}_1} = n_{\mathrm{CVB}_2} = n$ [22]. By two additional examples formed according to this approach, Fig. 5.5 gives an insight into the variety of 3d structured light fields carrying the spatially varying degree of non-separability. Here, $\ell = 2$ and (a) $n = 0$ or (b) $n = 1$ are chosen with

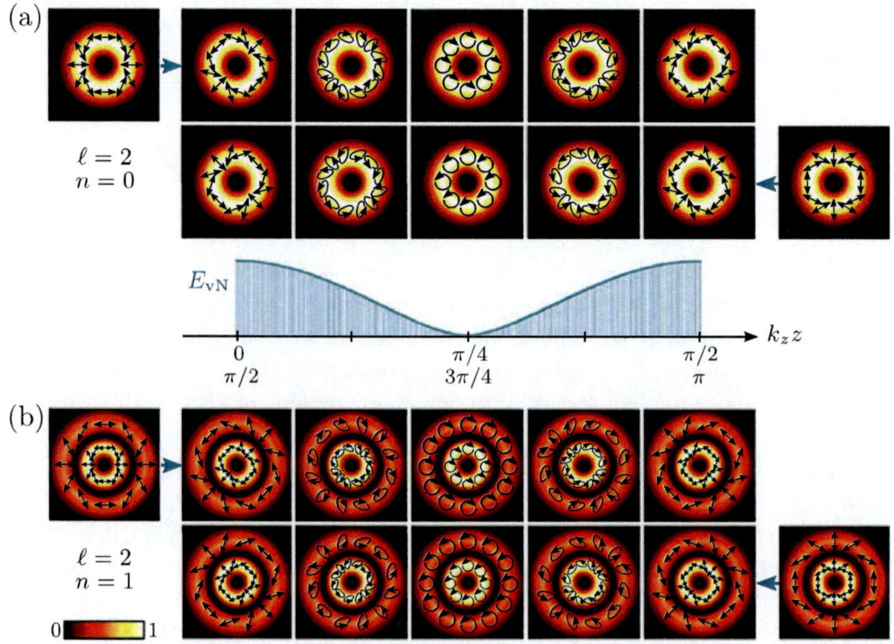

Fig. 5.5 The formation of entanglement beating by exemplary counter-propagating higher-order CVBs with $\ell = 2$ and **a** $n = 0$ or **b** $n = 1$. Shown are the normalized intensity and polarization structure (black arrows) for different propagation distances z (top row: $k_z z \in [0, \pi/2]$, bottom row: $k_z z \in [\pi/2, \pi]$; $\delta = 0$). Initial $\mathrm{CVB}_{1,2}$ are illustrated at the left ($+z$-propagation) and right ($-z$-propagation) edges. Blue: $E_{\mathrm{vN}}(z)$ curve. [Adapted with permission from Ref. [22], licensed under CC BY 4.0, https://creativecommons.org/licenses/by/4.0/]

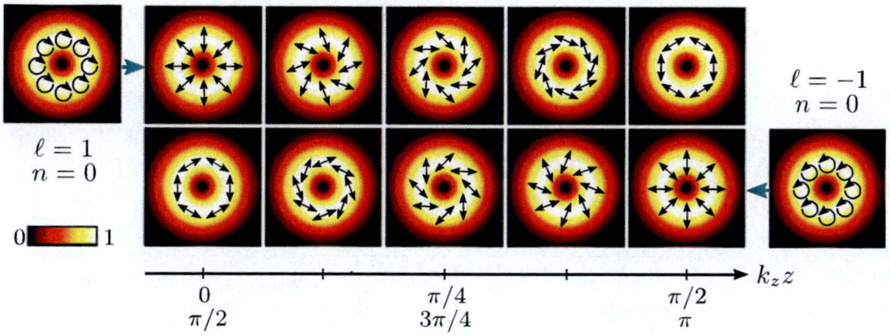

Fig. 5.6 Counter-propagating scalar LG modes of opposite topological charge $\ell = \pm 1$ and $n = 0$. Illustrated are the normalized intensity and polarization structure (black arrows) for different propagation distances z (top row: $k_z z \in [0, \pi/2]$, bottom row: $k_z z \in [\pi/2, \pi]$). Initial LG beams are sketched at the left ($+z$-propagation) and right ($-z$-propagation) edge. The degree of entanglement is constant with $E_{\mathrm{vN}} = 1$. [Adapted with permission from Ref. [22], licensed under CC BY 4.0, https://creativecommons.org/licenses/by/4.0/]

$\alpha_{\mathrm{CVB}_1} = 0$ and $\alpha_{\mathrm{CVB}_2} = \pi/2$. Shown are the states of polarization (black arrows) upon the normalized intensity distributions for different transverse (x, y)-planes with respective counter-propagating CVBs depicted at the left ($+z$-propagation) and right ($-z$-propagation) edges.

Crucially, for the implementation of the entanglement oscillation two counter-propagating non-separable states are required [22]. In contrast, if two fully separable states or, more precisely, two orthogonal scalar modes $|\Psi^{\pm}_{\mathrm{sc},1}\rangle$ and $|\Psi^{\mp}_{\mathrm{sc},2}\rangle$, are applied, the entanglement entropy stays constant upon propagation even though polarization varies longitudinally. This is easily demonstrated by the equation describing the respective superposition, namely

$$|\Psi_{\mathrm{3d,sc}}\rangle = \frac{1}{\sqrt{2}}\left(|\Psi^{\pm}_{\mathrm{sc},1}\rangle + |\Psi^{\mp}_{\mathrm{sc},2}\rangle\right) = \frac{1}{\sqrt{2}}\left(\mathrm{e}^{\pm \mathrm{i}k_z z}\,|\mathrm{LG}_{n,\ell}\rangle|R\rangle + \mathrm{e}^{\mp \mathrm{i}k_z z}\,|\mathrm{LG}_{n,-\ell}\rangle|L\rangle\right). \tag{5.24}$$

With $|u_R\rangle = \mathrm{e}^{\pm \mathrm{i}k_z z}\,|\mathrm{LG}_{n,\ell}\rangle$ and $|u_L\rangle = \mathrm{e}^{\mp \mathrm{i}k_z z}\,|\mathrm{LG}_{n,-\ell}\rangle$ one derives that $a = 1/2$, hence, $E_{\mathrm{vN}}(|\Psi_{\mathrm{3d,sc}}\rangle) = 1$ for all propagation distances z. An example of $|\Psi_{\mathrm{3d,sc}}\rangle$, being maximally non-separable, is presented in Fig. 5.6 with $\ell = \pm 1$ and $n = 0$.

Note that in the following, for simplicity but without the loss of generality, the study focuses on the creation of entanglement beating by lowest order CVBs, i.e., $\ell = 1$ and $n = 0$ for $|\Psi^{+}_{\mathrm{CVB}_1}\rangle$ and $|\Psi^{-}_{\mathrm{CVB}_2}\rangle$.

5.3.1.3 Paraxial Spin-Orbit Coupling in Free Space

As proven above, instead of a standing wave of intensity typically observed for counter-propagating longitudinally interfering scalar fields, a standing wave of entanglement is found for the combination of CVBs. This phenomena is based on the non-separability of the spatial and polarization DoF, i.e., the orbital (OAM) and spin angular momentum (SAM) of the initial fields. Within their counter-propagating combination, SAM and OAM couple in such a way that it causes the degree of entanglement to vary. Crucially, this effect represents a novel kind of SO coupling. Till now, SO coupling has solely been observed within non-paraxial light, paraxial light in anisotropic or inhomogeneous media, or through the spin-Hall effect of light at planar interfaces [37]. In the case at hand, SO coupling is realized in paraxial light propagating in free space [22].

For the $\ell = \pm 1$ subspace, the total angular momentum J_z is equal to zero for all propagation distances z. This can be derived from the respective equation [38]

$$J_z = \frac{\int \Im \left(\vec{\Psi}^* \cdot \partial_\varphi \vec{\Psi} + \vec{e}_z \cdot \vec{\Psi}^* \times \vec{\Psi} \right) \mathrm{d}\vec{r}}{\int \vec{\Psi}^* \cdot \vec{\Psi} \, \mathrm{d}\vec{r}}. \tag{5.25}$$

With $\vec{\Psi}$ as the classical analogon of $|\Psi_{3\mathrm{d}}\rangle$, $\vec{e}_z = [0,\ 0,\ 1]^T$ describing the longitudinal unit vector and the spatial position $\vec{r} = (r,\ \varphi,\ z)$ in polar coordinates, one obtains $J_z = 0$. Corresponding spin (S_z) and orbital (L_z) parts are calculated from [38]

$$S_z \propto \int \Im \left(\Psi_x^* \Psi_y - \Psi_y^* \Psi_x \right) \mathrm{d}A, \quad L_z \propto \int \Im \left(\Psi_x^* \partial_\varphi \Psi_x - \Psi_y^* \partial_\varphi \Psi_y \right) \mathrm{d}A \tag{5.26}$$

by the integral over the considered plane. Subscripts x and y refer to the horizontally and vertically polarized parts of $|\Psi_{3\mathrm{d}}\rangle = [\Psi_x,\ \Psi_y]^T$ (Jones vector in linear polarization basis).[4] By some algebra [22], one finds the proportionalities

$$S_z \propto \sin(2k_z z), \qquad L_z \propto -|\ell| \sin(2k_z z), \qquad J_z \propto (1 - |\ell|) \sin(2k_z z). \tag{5.27}$$

Therefore, while the spin/orbital part increases, the orbital/spin part decreases, i.e., they are counter-oscillating, showing paraxial SO coupling in free space [22]. Throughout this SO coupling, the total angular momentum is conserved with $J_z = 0$ for $\ell = \pm 1$ and $\forall z$.

5.3.2 *The Analysis of Entanglement Propagation Dynamics*

5.3.2.1 Experimental Implementation

In the following, the experimental study and, thus, verification of the theoretically described entanglement beating is performed. An intuitive approach for the implementation of the 3d structured field $|\Psi_{3\mathrm{d}}\rangle$, embedding the spatially varying degree

[4]Note that there is no longitudinal component Ψ_z considered, as paraxial fields are applied.

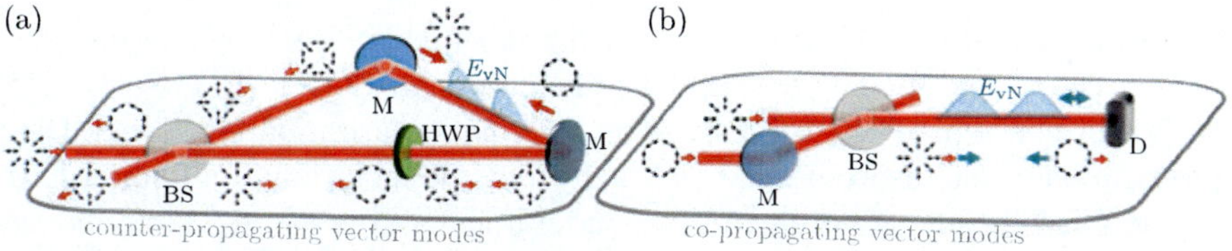

Fig. 5.7 Experimental approach of **a** counter- and **b** co-propagating modes for the realization of the 3d light field $|\Psi_{3d}\rangle$ and embedded local entanglement oscillation (blue: E_{vN} curve, red/blue arrows: physical/artifical propagation direction, black arrows: polarization of CVBs, BS: 50/50 beam splitter, M: mirror, HWP: half wave plate, D: detector). [Adapted with permission from Ref. [22], licensed under CC BY 4.0, https://creativecommons.org/licenses/by/4.0/]

of local entanglement, would be an interferometric one. An exemplary experimental system is proposed in Fig. 5.7a, in which a Sagnac interferometer is combined with a diagonally oriented HWP. Here, a single incident CVB, e.g., a radial mode shaped by a q-plate, is required, resulting in the on-axis counter-propagation of orthogonally polarized CVBs in each arm of the interferometer (black arrows: polarization states; red arrows: physical propagation direction). Hence, in all arms a standing wave of local entanglement is formed as exemplified by the blue E_{vN} curve in one arm.

Even though this approach is very straightforward, the experimental analysis of the light field $|\Psi_{3d}\rangle$ would be challenging: the insertion of any measuring device within the beam path would destroy the coupling within the light field, i.e., the oscillatory entanglement behavior could not be observed. To overcome this problem, an experimental design is proposed enabling the challenging quantification of the spatially varing degree of entanglement [22]. The technique relies on the co-propagating superposition of CVBs combined with artificial counter-propagation, as illustrated in Fig. 5.7b. Although physically, vector modes are co-propagating (red arrows), by the application of digital propagation via a SLM (cf. Sect. 3.3.3, Eq. (3.20)) the CVBs can be counter-propagated artificially (blue arrows) allowing for the study of $|\Psi_{3d}\rangle$.

Digital propagation is performed by the combination of the Fourier hologram representing $\mathcal{F}[E(x, y, 0)] \cdot \exp(ik_z z)$ and an optically Fourier transforming lens ($L_{\mathcal{F}}$). The hologram is encoded on a phase-only SLM enabling the formation of the desired light field $E(x, y, z)$ in the SLM's Fourier plane. Crucially, due to SLM-based technical restrictions, this method is only applicable to horizontally polarized scalar fields. Hence, for the implementation of digitally propagated CVBs an advanced method is demanded. For this purpose a holographic multiplexing approach for the realization of CVBs is designed [22, 39], which facilitates the generation of arbitrary vector modes as well as their independent manipulation even in global phase.

In order to holographically create a CVB which can be digitally propagated, holograms for the two underlying LG modes of opposite topological charge $\pm\ell$ are encoded on a SLM. Thereby, multiplexing is ensured by assigning each scalar LG mode to another spatial carrier frequency (blazed grating; cf. Sect. 2.4.1, Eq. (2.30)). The respective hologram creates the two LG beams spatially separated and horizontally polarized. In the next step, one of the two beams passes a HWP to create orthogonal states of polarization (horizontal and vertical). By superimposing the LG modes on-axis and sending the combined field through a QWP (fast axis at $\pm 45°$), a CVB of non-separable spatial shape and circular polarization is tailored. Digital propagation of the respective CVB is enabled by the inclusion of a Fourier lens in the combined beam path of the CVB, Fourier transforming the SLM into the plane of detection. Applying Fourier holograms of the CVB's LG modes with additional phase $\pm k_z z$ on the SLM, in the detection plane the CVB propagated in $\pm z$-direction is found.

As for the creation of $|\Psi_{3\mathrm{d}}\rangle$ two artificially counter-propagating CVBs need to be superimposed, four Fourier holograms of helical LG modes are multiplexed on a SLM and propagated accordingly. The experimental system, designed for this purpose, is sketched in Fig. 5.8a (blue box) with substeps I-IV visualized in Fig. 5.8b. First, scalar beams 1–4 are shaped by SLM_1, two of them pass a HWP for orthogonal polarization (I). Subsequently (II), beams 1 and 3 as well as 2 and 4 are combined for the creation of two CVBs. By superimposing the two CVBs and applying a QWP (QWP_1) the desired combination of CVB_1 (beam $1 + 3$) and CVB_2 (beam $2 + 4$) is formed (III). As beams 1–4 can be manipulated independently on SLM_1, in its Fourier plane (SLM_2) the underlying LG modes and, thus, the CVBs can be shifted independently in $\pm z$-direction (IV). Therefore, in the detection plane (SLM_2) the light field of interest $|\Psi_{3\mathrm{d}}\rangle$ can be studied. Crucially, this plane of observation stays static while the CVBs counter-propagate digitally. Moreover, the digital propagation also enables the implementation of the phase shift δ (or $\delta(t)$) and, thereby, the creation of a chosen degree of local entanglement at the observation plane, reminiscent of tractor beams.

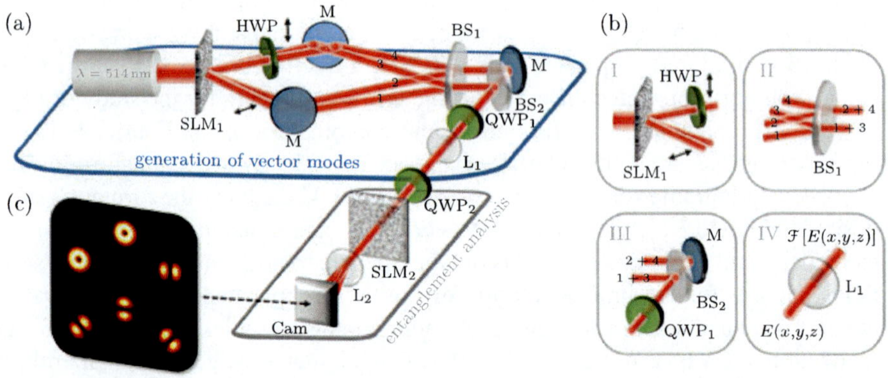

Fig. 5.8 Sketch of the applied experimental system for implementing and analyzing $|\Psi_{3\mathrm{d}}\rangle$. **a** Generation of physically co-propagating CVBs (blue box), counter-propagated artificially, with experimental steps shown in **b**. **c** System for entanglement analysis (grey box). (SLM: spatial light modulator, HWP/QWP: half/quarter wave plate, M: mirror, BS: 50/50 beam splitter, Cam: camera) [Adapted with permission from Ref. [22], licensed under CC BY 4.0, https://creativecommons.org/licenses/by/4.0/]

To study the realized 3d structured field and its degree of entanglement, two approaches are applied in the following [22]. On the one hand, a camera is positioned in the detection plane (Fourier plane of SLM_1) to study intensity structures upon artificial counter-propagation. On the other hand, the degree of entanglement or, more precisely, the entanglement entropy is quantified in the detection plane, as depicted in Fig. 5.8c. The local entanglement analysis is performed as described in Sect. 5.1.2 with polarization/spatial projections realized by QWP_2/SLM_2. SLM_2 is positioned in the Fourier plane of SLM_1 (Fourier relation through L_1) and displays the six decoding holograms in a multiplexed fashion. The on-axis intensity values I_{uv} are detected by a camera seeing the far field of SLM_2 (Fourier lens L_2). To determine the degree of non-separability for different z-distances within $|\Psi_{3d}\rangle$ and, thereby, to verify that it follows the entanglement dynamics predicted in Eq. (5.21), the analysis is combined with digital propagation.

5.3.2.2 Counter-Fluctuating Polarization Contributions

Following the outlined approach, the superposition of artificially counter-propagating first-order vector beams, namely, a radial and azimuthal CVB is performed. Of course, the system is not limited to $\ell = 1$ and $n = 0$ but could also be applied for any OAM subspace with variable n. As visible in Eq. (5.20), the field $|\Psi_{3d}\rangle$, artificially realized in the experiment, is constituted of $|R\rangle$ and $|L\rangle$ parts, each including two counter-propagating LG modes of the same helicity $\pm\ell$. Counter-propagation of respective LG modes per polarization is realized by SLM_1, encoding $\exp(\pm i[k_z z + \delta])$. In the case at hand, $\delta = -\pi/4$ is chosen. To study the counter-propagating behavior of $|\Psi_{3d}\rangle$, the intensity profile of the $|R\rangle$ as well as $|L\rangle$ part is recorded separately in the detection plane for different propagation distances. To do so, beam 3 and 4 or 1 and 2 are shut for observing purely right- or left-handed polarized states, respectively. Results are presented in Fig. 5.9. Here, in Fig. 5.9a the polarization configuration of $|\Psi_{3d}\rangle$ per z-distances is sketched with the according curve of the entanglement degree in the background (blue). For comparison reasons, in Fig. 5.9b the calculated transverse intensity distribution of $|\Psi_{3d}\rangle$ is given, assuming a horizontally aligned polarizer in front of the camera. These intensity images reflect the polarization structures shown in Fig. 5.9a. Experimentally recorded intensity landscapes for the left- (beam $3 + 4$) and right-circularly (beam $1 + 2$) polarized parts are presented in Fig. 5.9c and d, respectively, for holographically adjusted values of $k_z z + \delta \in [0, \pi]$. For both, $|R\rangle$ as well as $|L\rangle$ components, a sinusoidal intensity variation is found, revealing longitudinal interference between counter-propagating LG modes for each polarization component. Comparing Fig. 5.9c and d, one sees that the found variation for $|R\rangle$ and $|L\rangle$ is out of phase, such that $|R\rangle$ carries maximum intensity when the intensity of $|L\rangle$ contributions is minimal and vice versa. Note that this behavior is due to the phase shift $\alpha_{CVB_{1,2}}$, which is introduced to create the polarization orthogonality of CVBs (cf. Eqs. (5.18), (5.19)). Crucially, the counter-fluctuating behavior gives proof of the oscillation between pure scalar (separable) and vector (non-separable) states within the 3d light field $|\Psi_{3d}\rangle$: In case $|R\rangle$ and $|L\rangle$ components

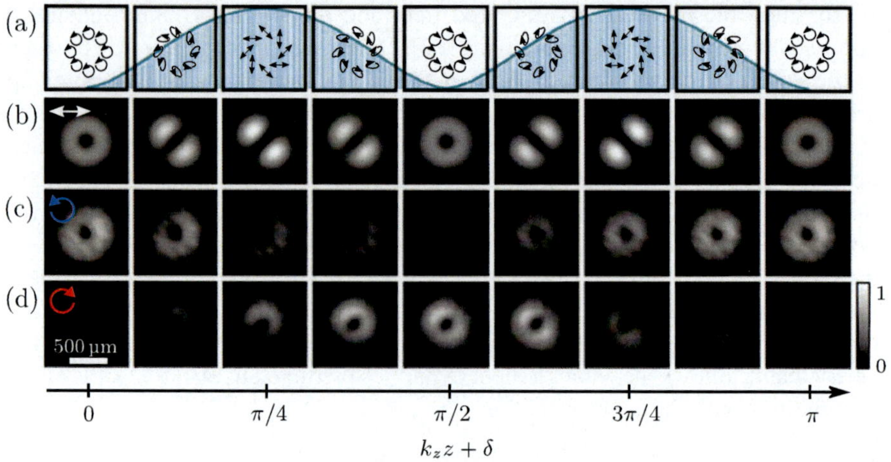

Fig. 5.9 Observance of counter-fluctuating polarization contributions in $|\Psi_{3d}\rangle$ depending on $k_z z + \delta \in [0,\ \pi]$, $\delta = -\pi/4$. **a** Sketch of transverse polarization configuration and **b** intensity profiles of $|\Psi_{3d}\rangle$ considering a horizontally aligned polarizer in front of the camera (calculations). **c/d** Experimentally recorded intensity profiles of $|R\rangle/|L\rangle$ polarized parts. [Adapted with permission from Ref. [22], licensed under CC BY 4.0, https://creativecommons.org/licenses/by/4.0/]

are of equal intensity, $|\Psi_{3d}\rangle$ represents a pure vector state with the maximal degree of local entanglement $E_{vN}(|\Psi_{3d}\rangle, z) = 1$ observed at $k_z z + \delta = \{\pi/4,\ 3\pi/4\}$. In contrast, a pure scalar mode ($E_{vN}(|\Psi_{3d}\rangle, z) = 0$) is found for $k_z z + \delta = \{0,\ \pi/2,\ \pi\}$. Here, the $|R\rangle$ ($|L\rangle$) polarized parts are of maximal intensity while $|L\rangle$ ($|R\rangle$) polarized ones disappear, so that the total field $|\Psi_{3d}\rangle$ is solely constituted of $|R\rangle$ ($|L\rangle$) parts and, hence, is fully separable. Between the extreme cases of entire separability and maximal non-separability, a smooth transition is found as visible in Fig. 5.9a and b.

5.3.2.3 Quantification of the Local Entanglement Degree

In addition to its qualitative verification by recorded intensity profiles, here, the oscillation of local entanglement within $|\Psi_{3d}\rangle$ is proven quantitatively [22]. For this purpose, the entanglement entropy $E_{vN}(|\Psi_{3d}\rangle, z)$ is measured as a function of $k_z z + \delta \in [0,\ \pi]$ ($\delta = -\pi/4$) by means of artificial counter-propagation. Corresponding to this measurement procedure, Fig. 5.10 gives three examples of experimental I_{uv} results and respective E_{vN} values. Here, in Fig. 5.10a examples of typically observed intensity images are presented for both polarization projections. White crosses mark the on-axis positions for the determination of respective intensity values I_{uv}. Presented images correspond to the scalar state illustrated in Fig. 5.10b. As clearly visible within presented intensity values (arranged according to Table 5.1), the field is purely left-circularly polarized with an OAM of charge $\ell = -1$. Accordingly, based on these values the entanglement entropy is computed to be $E_{vN} = 0.01$. In Fig. 5.10c and d the intensity values of two additional examples are given, pre-

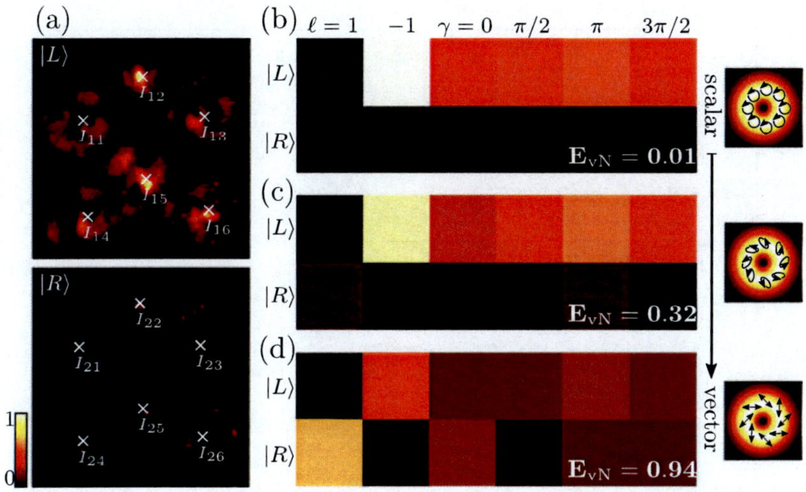

Fig. 5.10 Exemplary measurement results of the entanglement analysis for a **a/b** scalar, **c** semi-vector and **d** vector state within $|\Psi_{3d}\rangle$. **a** Typically recorded intensity images and **b–d** intensity values I_{uv} arranged according to Table 5.1 are shown. Determined entanglement entropy E_{vN} is given within subfigures. [Adapted with permission from Ref. [22], licensed under CC BY 4.0, https://creativecommons.org/licenses/by/4.0/]

senting a semi-vector and vector state, respectively. Matching the expectations for these kinds of states, experimental measurements result in $E_{vN} = 0.32$ (semi-vector) and $E_{vN} = 0.94$ (vector).

In Fig. 5.11a the complete set of experimentally determined E_{vN} values (colored circles) are presented as a function of propagation distance z ($k_z z + \delta \in [0, \pi]$, $\delta = -\pi/4$) [22]. Additionally, Fig. 5.11b depicts respective normalized intensity values I_R (black filled diamonds, red fit) and I_L (black hollow diamonds, blue fit) of the $|R\rangle$ and $|L\rangle$ parts of $|\Psi_{3d}\rangle$, respectively. Error bars correspond to SLM flickering affecting $k_z z + \delta$ ($\pm\pi/16$) and inaccuracies within the experimental method/system causing errors within E_{vN} (± 0.05) and $I_{R,L}$ (± 0.03).

To evince the typically invariant degree of non-separability of pure vector states upon propagation, a radially polarized CVB is exemplarily analyzed. Results are presented by black triangles (filled white) in Fig. 5.11a. Fulfilling the expectation, the CVB has an approximately maximal degree of entanglement $E_{vN} = 1$ for all propagation distances. In contrast, experimental results for the light field $|\Psi_{3d}\rangle$ (Eq. (5.20)) verify the oscillation between pure scalar and vector fields. Respective data points (black circles) are colored according to the ratio of $|L\rangle$ (blue) and $|R\rangle$ (red) components (cf. colorbar in Fig. 5.11a) emphasizing the oscillatory entanglement dynamics. Additionally, insets of spatial modes with polarization vectors indicate the state of light at selected positions. The fitting curve (black dashed line) following Eq. (5.10) with $k_z z$ being replaced by $k_z z + \delta'$ reveal the perfect match between experimental results and theoretical predictions. With a value of $\delta' = -0.71$ the fitting parameter almost agrees with the chosen setting of $\delta = -\pi/4$.

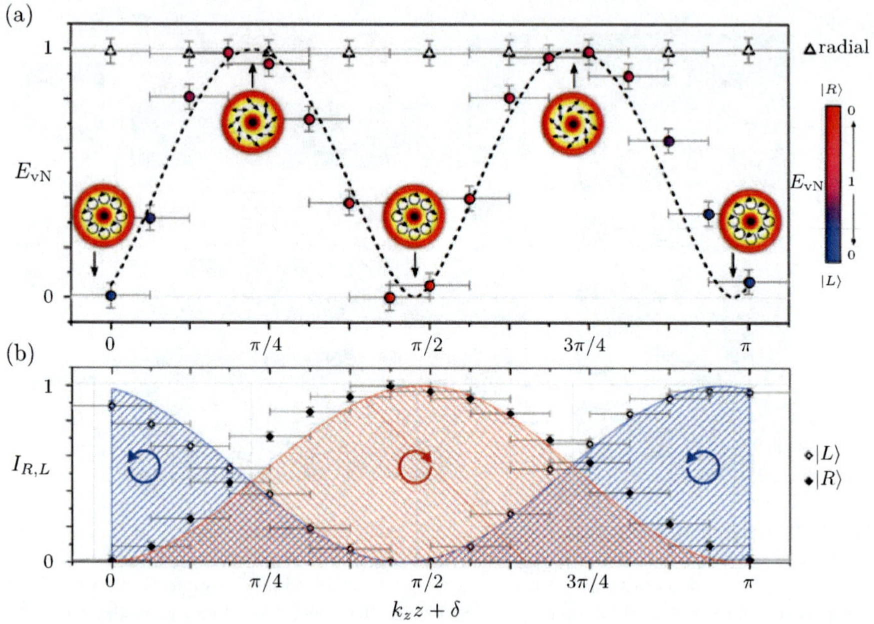

Fig. 5.11 Experimental study of entanglement propagation dynamics. **a** Local entanglement analysis of a radial CVB (black triangles) and the light field $|\Psi_{3\text{d}}\rangle$ (back circles) as a function of $k_z z + \delta \in [0, \pi]$, $\delta = -\pi/4$. Data points of the latter are filled according to the ratio of $|L\rangle$ (blue) and $|R\rangle$ (red) components (see colorbar). Insets show exemplary states within $|\Psi_{3\text{d}}\rangle$. **b** Counter-fluctuating normalized intensity values $I_{R,L}$ of $|L\rangle$ (black hollow diamonds, blue fit) and $|R\rangle$ (black filled diamonds, red fit) components. [Adapted with permission from Ref. [22], licensed under CC BY 4.0, https://creativecommons.org/licenses/by/4.0/]

Confirming previous results, Fig. 5.11b shows the variation in ratio between $|L\rangle$ and $|R\rangle$ components, reflecting the entanglement propagation dynamics visualized in Fig. 5.11a. Mirroring the deviation between δ' and δ, the z-position of the maximum and minimum of $|L\rangle$ ($|R\rangle$) and $|R\rangle$ ($|L\rangle$), respectively, is slightly shifted to each other. However, the results evince that the adjustment of δ allow for the transport of a chosen degree of local entanglement to a predefined position in space.

5.3.3 Discussion and Future Perspectives

Presented results prove the ability to engineer a 3d structured vectorial light field whose degree of local entanglement, i.e., non-separability varies upon free space propagation, oscillating between fully vector and pure scalar states. Crucially, this behavior is due to paraxial SO interaction in free space—a novel kind of SO coupling [22]. Although experimental studies were restricted to CVBs of first order with

$\ell = \pm 1$, the outlined scenario is more general and applicable to any set of counter-propagating vector modes with a judicious choice of DoFs. Beyond, the phenomenon observed under unitary conditions is not limited to coherent light. The same results are obtained for a single photon being locally entangled in its internal DoFs with neither theory nor experiments differentiating from the coherent case. Hence, cutting-edge questions as to the notion of local or classical entanglement and its propagation dynamics are addressed.

Further, the experimental approach for realization and analysis of the studied field is worth being highlighted, as it presents an innovative technique to measure counter-propagating CVBs without destroying their interaction [22]. Intuitively, the respective implementation of the 3d structured field could be performed by an interferometric method, however, corresponding techniques do not allow for the experimental verification of the spatially varying non-separability. The alternatively proposed powerful approach to digitally generate and artificial counter-propagate respective CVBs does not only facilitate the quantification of entanglement beating, but even the transport of a chosen degree of entanglement to a predefined z-position. The simple adaption of a hologram enables the formation of any states embedded in $|\Psi_{3d}\rangle$ at the observer's fixed position.

Moreover, the ability to dynamically adapt the state of light at the observer's position or transport a selected state to chosen position in space paves the way to the enhancement of several noteworthy applications [22], as summarized in Fig. 5.12. For instance, rapidly changing the state of light from a radial CVB to a donut shaped beam

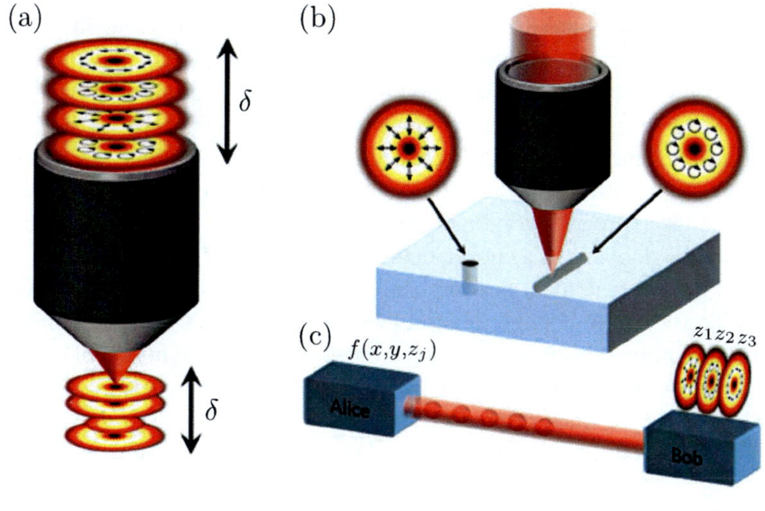

Fig. 5.12 Exemplary applications of artifically counter-propagated CVBs of orthogonal polarization. **a** Dynamic adjustment of focal light fields by phase shift δ for, e.g., STED microscopy, optical manipulation, or **b** laser material machining (focused radial CVB for drilling, scalar circularly polarized mode for cutting). **c** Sketch of innovative QKD approach for Alice delivering states to Bob. [Adapted with permission from Ref. [22], licensed under CC BY 4.0, https://creativecommons.org/licenses/by/4.0/]

enables the dynamic switching of focal light fields from a tight spot to a donut (a), which perfectly fulfills the requirement for STED microscopy [40–42]. Furthermore, respective rapid change will be of clear benefit for (b) laser material processing [43–45]. In this case a radially polarized field is favourable for drilling, whereas for cutting, the light field would be dynamically adapted to a circularly polarized beam. Additionally, presented results open new possibilities for QKD approaches using a prepare-and-measure BB84 QKD protocol [46] with respective scalar and vector modes as orthogonal and mutually unbiased bases [28]. Here, Alice is able to deliver chosen states to Bob by simply adjusting holograms.

Going from applied back to fundamental classical optics, the realized 3d field $|\Psi_{3d}\rangle$ is of special interest considering singular optics. Obviously, within the scalar states appearing for specific z-distances a scalar phase singularity characterized by ℓ is embedded. In contrast, in pure vector states a non-generic V-point (idealized case), i.e., a vectorial singularity of index σ_{12}, is found. In between these two states of light, a transition between these two types of singularities will be observed. This transition gives new insights into not yet observed singularity dynamics and addresses topical questions, e.g., on singularity index conservation if two classes of singularities are engaged.

5.3.4 Conclusion

This section demonstrates the ability to create an oscillating degree of local entanglement during propagation, although the respective medium (free space) is considered as unitary, i.e., the entanglement should be invariant. Additionally, results evince paraxial SO coupling in free space, hence a not yet realized form of SO interaction. Entangling the spatial shape and polarization DoF, the phenomenon was observed for counter-propagating CVBs, realized and studied by an innovative experimental approach. This approach additionally enables the first tractor beam of entanglement, facilitating the delivery of a chosen degree of non-separability to some target plane. Crucially, the entanglement beating is not only true for classical coherent light, but can also be formed at a single photon level with entangled internal DoFs. In total, the presented approach addresses topical questions about the notion of entanglement dynamics and conservation of singularity indices, present a new type of SO coupling, and offers a tool for the advancement of various application for which the dynamic holographic availability of vector and scalar states of light is of huge benefit.

References

1. Schrödinger E (1935) Die gegenwärtige Situation in der Quantenmechanik. Naturwissenschaften 23: 807, 823, 844
2. Monroe C et al (1966) A "Schrödinger Cat" superposition state of an atom. Science 272:1131
3. Einstein A, Podolsky B, Rosen N (1935) Can quantum-mechanical description of physical reality be considered complete? Phys Rev 47:777
4. Bell JS (1964) On the Einstein Podolsky Rosen paradox. Phys Physique Fizika 1:195
5. Ekert AK (1991) Quantum cryptography based on Bell's theorem. Phys Rev Lett 67:661
6. Deutsch D (1985) Quantum theory, the Church-Turing principle and the universal quantum computer. Proc R Soc Lond A 400:97
7. Bennett CH (1995) Quantum information and computation. Phy Today 48:24
8. Aiello A et al (2015) Quantum- like nonseparable structures in optical beams. New J Phys 17:043024
9. Ndagano B et al (2017) Characterizing quantum channels with non-separable states of classical light. Nat Phys 13:397
10. Guzman-Silva D et al (2016) Demonstration of local teleportation using classical entanglement. Laser Photon Rev 10:317
11. Töppel F et al (2014) Classical entanglement in polarization metrology. New J Phys 16:073019
12. Borges C et al (2010) Bell-like inequality for the spin-orbit separability of a laser beam. Phys Rev A 82:033833
13. Karimi E et al (2010) Spin-orbit hybrid entanglement of photons and quantum contextuality. Phys Rev A 82:022115
14. Pereira L, Khoury A, Dechoum K (2014) Quantum and classical separability of spin-orbit laser modes. Phys Rev A 90:053842
15. Rosales-Guzmán C, Ndagano B, Forbes A (2018) A review of complex vector light fields and their applications. J Opt 20:123001
16. Souza C et al (2008) Quantum key distribution without a shared reference frame. Phys Rev A 77:032345
17. Milione G et al (2008) Using the nonseparability of vector beams to encode information for optical communication. Opt Lett 40: 4887)
18. Li P, Wang B, Zhang X (2016) High-dimensional encoding based on classical nonseparability. Opt Express 24:15143
19. Milione G et al (2015) 4x 20 Gbit/s mode division multiplexing over free space using vector modes and a q-plate mode (de)multiplexer. Opt Lett 40:1980
20. Otte E et al (2018) Recovery of nonseparability in self-healing vector Bessel beams. Phys Rev A 98:053818
21. Nape I et al (2018) Self-healing high-dimensional quantum key distribution using hybrid spin-orbit Bessel states. Opt Express 26:26946
22. Otte E et al (2018) Entanglement beating in free space through spin-orbit coupling. Light Sci Appl 7:18009
23. McLaren M, Konrad T, Forbes A (2015) Measuring the nonseparability of vector vortex beams. Phys Rev A 92: 023833)
24. Hill S, Wootters WK (1997) Entanglement of a pair of quantum bits. Phys Rev Lett 78:5022
25. Wootters WK (2017) Entanglement of formation and concurrence. Q Inf Comput 1:27
26. Wootters WK (1998) Entanglement of formation of an arbitrary state of two qubits. Phys Rev Lett 80:2245
27. Ndagano B et al (2016) Beam quality measure for vector beams. Opt Lett 41:3407
28. Ndagano B et al (2017) A deterministic detector for vector vortex states. Sci Rep 7:13882
29. Clauser JF et al Proposed experiment to test local hidden-variable theories'. Phys. Rev. Lett. 23 (1969), 880 (cit. on p. 123)
30. Kagalwala KH et al (2013) Bell's measure in classical optical coherence. Nat Photon 7:72
31. Cirel'son BS (1980) Quantum generalizations of Bell's inequality. Lett Math Phys 4:93

32. Mintert F, Viviescas C, Buchleitner A (2009) Basic concepts of entangled states. In: Entanglement and decoherence. Springer, pp 61–86
33. Novitsky A, Qiu C-W, Wang H (2011) Single gradientless light beam drags particles as tractor beams. Phys Rev Lett 107:203601
34. Brzobohatý O et al (2013) Experimental demonstration of optical transport, sorting and self-arrangement using a 'tractor beam'. Nat Photon 7:123
35. Ruffner DB, Grier DG (2012) Optical conveyors: a class of active tractor beams. Phys Rev Lett 109:163903
36. Gorlach AA et al (2017) Matter-wave tractor beams. Phys Rev Lett 118:180401
37. Bliokh KY et al (2015) Spin-orbit interactions of light. Nat Photon 9:796
38. Berry M, Jeffrey M, Mansuripu M (2005) Orbital and spin angular momentum in conical diffraction. J Opt A: Pure Appl Opt 7:685
39. Rosales-Guzmán C et al (2017) Multiplexing 200 spatial modes with a single hologram. J Opt 19:113501
40. Hell SW, Wichmann J (1994) Breaking the diffraction resolution limit by stimulated emission: stimulated-emission-depletion fluorescence microscopy. Opt Lett 19:780
41. Török P, Munro P (2004) The use of Gauss-Laguerre vector beams in STED microscopy. Opt Express 12:3605
42. Hao X et al (2010) Effects of polarization on the de-excitation dark focal spot in STED microscopy. J Opt 12:115707
43. Nivas JJ et al (2017) Surface Structuring with Polarization-Singular Femtosecond Laser Beams Generated by a q-plate. Sci Rep 7:42142
44. Niziev VG, Nesterov AV (1999) Influence of beam polarization on laser cutting efficiency. J Phys D: Appl Phys 32:1455
45. Meier M, Romano V, Feurer T (2007) Material processing with pulsed radially and azimuthally polarized laser radiation. Appl Phys A 86:329
46. Bennett CH, Brassard G (1989) Experimental quantum cryptography: the dawn of a new era for quantum cryptography: the experimental prototype is working. ACM Sigact News 20:78

Chapter 6
Summary and Perspectives

This thesis presents topical studies in the field of structured singular light fields, which shed new light on fundamental optics and contribute to the future of applied optics. Beginning with vectorial light sculpted in 2d space, via 3d tailored fields, through to sophisticated 4d structures, the ability to customize vectorial light fields as well as embedded optical singularities in their different properties was examined. Beyond, enhancing the rich phenomenology of complex singular light, the quantum attribute of local entanglement was transferred to and studied in classical fields. The performed investigations pave the way to advanced applications, e.g., in optical micromanipulation, the implementation of 4d materials, or novel quantum key distribution (QKD) approaches.

In the first chapter, the fundamental and experimental basis for the subsequent studies was provided. On the one hand, the concept of singularities in general was introduced, on the other hand the fundamentals of singular light being structured in amplitude, phase, as well as polarization were presented. On this fundamental basis, Chap. 2 outlines different experimental customization techniques for the realization of singular light fields, especially vectorial ones. Established methods were presented, whereby their not yet fully exploited potential was elucidated by the innovative realization of tailored singular polarization and spin flow density (SFD) structures in 2d space. Crucially, the study of singular properties within these 2d structures gave new insights into the properties of vectorial singularities in close vicinity and the on-demand formation of optical SFD singularities. Hence, respective findings contribute to fundamental singular optics, but also to applied physics, as tailored SFD configurations are of particular interest for the optical tweezers community. Additionally, joining established structured light techniques, a novel customization method was developed for sculpting fully-structured light, i.e., light being spatially shaped in amplitude, phase as well as polarization. This spatial multiplexing-based technique is an innovative advancement for experimental systems, as holographic

E. Otte, *Structured Singular Light Fields*, Springer Theses,
https://doi.org/10.1007/978-3-030-63715-6_6

optical tweezers or laser material machining systems, which desire the inclusion of additional polarization modification in their scalar modulation system.

Going beyond presented 2d structures, Chap. 3 features the ability to shape vectorial light fields and its optical singularities in 3d space. This chapter reveals not yet known properties of optical singularities and their interaction characteristics upon propagation. Here, not only individual singularities were examined, but even extended networks of polarization singularities were shaped on demand. For this purpose, applying incoherent superposition, a novel class of vectorial fields was proposed, namely, Ince-Gaussian vector modes (IVMs). These modes include complex singularity networks (SNs) whose properties can be shaped by the adaption of mode indices. Interestingly, created SNs mimic the form singularities naturally occur in—arrangements of multiple singular points in a joint light field. Crucially, IVMs can be tailored to propagate structurally stably as well as unstably in 3d space, whereas the overall configuration of the embedded SNs is always conserved even under turbulence. Due to their stability characteristics these modes are of specific interest for free space data communication with a significantly increased number of degrees of freedom (DoFs).

The study of the stability characteristics of optical singularities was further extended by the investigation of interaction properties between different singularity classes. For this purpose, non-generic scalar and vectorial singular points are combined, meandering through 3d space and forming "singularity explosions". The numerical examination revealed the adaptability of propagation dynamics by the transfer of unfolding properties of scalar to vectorial singularities. Another highlight of this section is the study of multiple V-point and phase singularities of opposite index and charge, respectively, in very close vicinity. Here, it was demonstrated that, although vectorial/ scalar singularities of opposite index/ charge may annihilate each other in a specific (x, y)-plane, the splitting of each singularity in its generic counterparts can be tailored by structured light techniques. Further, the numerical investigations evinced the generic form of line singularities in complex light fields in 3d space: while non-generic C-lines in 2d space unfold upon 3d propagation, generic L-lines are conserved creating L-surfaces in 3d space.

Additionally, extending the ability to customize propagation dynamics of structured vectorial light including its singularities in 3d space, valuable properties of 3d structured scalar fields were transferred to vectorial beams. More precisely, non-diffracting, self-healing and self-imaging vectorial light fields were shaped embedding individual singularities or complete SNs. First, vector Bessel-Gaussian (vBG) modes were experimentally modulated and examined by the combination of a spatial light modulator (SLM), a q- and different wave plate(s). Not only the propagation invariant vectorial beam profile was evinced, but also the feature of self-reconstruction of these fields was proven by an all-digital SLM-based approach. Significantly enlarging the class of propagation invariant vectorial fields, vectorial discrete non-diffracting beams were proposed and experimentally realized by far-field construction joint with an amplitude, phase and polarization modulation system. Crucially, these fields embed vectorial SNs, which also propagate spatially invariant. Proving the not yet observed combination of transverse and longitudinal customization of fully-structured fields and polarization singularities, the far-field approach was extended to form vectorial self-imaging light fields. The self-replicating characteristic of these fields was proven exemplarily by combining the far-field construction and digital propagation. Beyond, for the first time, the 3d shaping of singularity propagation dynamics was evinced, including the creation and annihilation of C- and V-point singularities.

Chapter 4 paved the way to include a fourth dimension to structured light fields: By tightly focusing modulated paraxial beams, the creation of non-paraxial complex light fields is enabled including 3d instead of standard 2d polarization. Crucially, by tailoring the incident field in phase and polarization, the focal 4d field was proven to be customized on demand in its 3d spatial shape as well as the ratio of transverse and longitudinal electric field components. As customization tools the vectorial and scalar singularity indices σ_{12} and ℓ, respectively, of the incident fields were proposed. Focal intensity landscapes as bright flowers and dark stars were realized by pure polarization modulation. By the addition of phase vortices, another DoF for the customization was provided facilitating not only tuning the longitudinal polarization components but even twisting intensity structures in the 3d focal volume. The significant benefit of respective tailored 4d fields was proven by trapping and orienting zeolite-L nano-containers, which may particularly profit from the on-demand 3d polarization structure.

For the effective implementation of structured 4d fields their experimental identification represents a major challenge. The detailed study of non-paraxial field properties is impeded by their nano-scale complexity and, especially, the 3d polarization nature. Tackling this topical difficulty, a pathbreaking nano-tomographic approach was suggested, applying a 4d material as sensor. More precisely, fluorescent self-assembled monolayers were proposed as optical nano-sensors, which react sensitive to focal amplitude, phase as well as 3d polarization at the nano-meter scale in all three spatial dimensions. Beyond, monolayers facilitate the till now impossible single-shot identification of typically invisible non-paraxial field properties. By proving this feature theoretically as well as experimentally, the effective experimental study and subsequent application of 4d light fields is finally enabled.

Considering the application of 4d focal fields in the area of 4d materials or assembly of polarization sensitive objects, the embedded 3d polarization topology and included singularities are of specific interest. It was proven that by the customization tool of tailoring the paraxial input field, the number and position of generic L- and C-singularities in the focal plane can be chosen on demand by indices σ_{12} and ℓ. Beyond, this tool even provides the opportunity to shape the 3d topology surrounding the optical axis or off-axis singularities in the focal plane. As proven numerically, the major axes around the optical axis may form optical cones or $|\sigma_{12} - 2|$ times twisted ribbons. As another highlight within this chapter, it was even demonstrated that the 3d polarization can be tailored in such a way, that respective major axes create a not yet observed array of optical Möbius strips—each strip found around tailored off-axis C-singularities.

Opening new perspectives for applied optics, in Chap. 5 valuable quantum properties are fused with classical ones by the study of local entanglement or non-separability in tailored classical light. As outlined within this chapter, the respective fusion has already paved the way to, e.g., novel QKD approaches. While the local entanglement of spatial shape and polarization in cylindrical vector beams (CVBs) is well-established, adapting the spatial modes including its radial DoF has not been considered so far. Within this chapter, by choosing non-diffracting higher-order Bessel modes as basis, classically entangled vBG modes were created and studied experimentally with respect to their entanglement properties. Crucially, it was demonstrated that the classical property of self-healing can be transferred to the degree of entanglement within obstructed vBG modes. Interestingly, the self-reconstruction is typically attributed to the radial component of the light field, which is totally separable in the considered case. Hence, it was proven that by adapting this radial component, local entanglement can be made more resilient to decay from obstructions. This ability has already proven its benefit for QKD through obstacles.

In the last section of this thesis, the ability to shape local entanglement beating upon free space propagation was demonstrated in a 3d structured field, even though within the respective unitary medium the entanglement degree should not change. The found phenomenon was created by a not yet observed kind of spin-orbit (SO) interaction, namely, paraxial SO coupling in free space. This novel kind of coupling was formed by counter-propagation of two orthogonally polarized and locally entangled CVBs. As theoretically proven, within the respective on-axis superposition the degree of entanglement varies as a function of propagation distance z. Interestingly, by a simple phase adjustment the transport of a chosen degree of entanglement, i.e., of a specific plane of the overall light field, is facilitated. Hence, the first tractor beam of entanglement was proposed. By an innovative experimental approach, it was possible to evince the theoretically outlined scenario of entanglement oscillation and, thus, of paraxial free space SO coupling. Addressing topical questions about the notion of local entanglement dynamics, the proven entanglement beating is not only true for classical coherent light, but also at a single photon level with entangled internal DoFs. Beyond, the experimental approach offers a holographic means for the dynamic availability of vector and scalar states of light being a huge benefit for applications as QKD, laser material machining, optical micromanipulation or imaging.

6.1 Future Perspectives

As partially indicated above, various results of this thesis give rise to future projects and, thereby, open up new perspectives for fundamental and applied research.

Performed studies make a significant contribution to the field of fundamental singular optics. Not yet known properties of optical scalar and vectorial singularities are revealed, including, for instance, their interaction characteristics upon propagation in 3d space. Beyond, the on-demand formation of SNs in tailored fields mimic the form singularities naturally occur in, i.e., arrangements of multiple singular points. Furthermore, singularities were also studied within 4d light fields, giving new insights into the properties of singularities in the non-paraxial regime. Crucially, the techniques applied for shaping singularities in 2d, 3d, and 4d structured fields facilitate many more studies on optical singularities: dynamic methods as combining non-generic scalar and vectorial singularities or shaping the input field to customize the resulting focal 4d field may be used to create novel singularity configurations. For instance, paraxial and non-paraxial vectorial singularity knots or networks of knotted structures could be sculpted, going beyond cutting-edge polarization knots realized only recently [1]. Moreover, the approach of counter-propagating light fields enabling the oscillation between scalar and vector states of light facilitate the not yet performed examination of optical singularity transformations from scalar to vectorial singularities. This transformation gives rise to questions about the singularity index conservation rule and, thus, fundamental characteristics of singularities. Beyond, the counter-propagation technique may be advanced to various different spatial modes, far beyond CVBs, enabling novel complex 3d shaped light fields including sophisticated singularity configurations which have not been studied so far.

Considering future applications, the transverse as well as longitudinal customization of vectorial fields opens new perspectives, profiting from different tailored properties: structural stability, nano-scale complexity in 2d or 3d space and the 2d or even 3d polarization nature. For instance, due to their structural stability characteristics, novel fully-structured light fields as IVMs are of specific interest for research fields as optical free space data communication. Beyond, innovative amplitude, phase and, especially, polarization structures will be of huge benefit for high-resolution imaging, material machining, or optical micromanipulation. In this context, in particular non-diffracting and self-imaging vectorial fields, 3d structures created by counter-propagation, or tightly focused beams, customizable at nano-meter resolution, are of interest. Considering the latter, an exemplary application has been demonstrated within this thesis by trapping and orienting cylindrical particles. Beyond, the applied 4d fields have the potential to additionally selectively excite loaded molecules. This characteristic may pave the way to their implementation, e.g., in medical application profiting from selective excitation of pharmaceuticals.

Finally enabling the effective implementation of 4d fields, a special highlight of promising future represents the nano-tomographic technique based on self-assembled monolayers, i.e., a 4d material. The presented capability of monolayers is urgently demanded for the experimental study and subsequent implementation of 4d focal fields in above mentioned applications. In particular, experimentally analyzed 4d fields and imparted 3d topological structures pave the way to the optimal implementation of 4d materials and the precision assembly of phase and polarization sensitive objects. Beyond, as described in detail in Sec. 4.3.3, by advanced monolayers of defined orientation, the full electric field reconstruction can be facilitated in the future. This ability may enable the experimental verification of complex 4d fields till now only described theoretically, e.g., non-paraxial fields embedding polarization singularity knots within its transverse as well as longitudinal field components in 3d space [2].

Also the study of local entanglement in tailored vectorial light fields lead to findings which will advance applied optics. On the one hand, it was demonstrated that shaping the separable radial component of spatial modes may be used to assign beneficial classical properties to local entanglement and respective applications as QKD, as self-healing. Furthermore, the ability to create a 3d structured field carrying entanglement oscillation bears the potential to develop novel QKD schemes using the property to dynamically deliver scalar or vector modes to a receiver. This delivery is also of interest for classical applications as optical trapping, for which the embedded novel kind of SO coupling may be of specific benefit to orbit or spin particles in chosen z-planes. Additionally, the concept of counter-propagation of structured light fields, e.g., higher-order CVBs or just scalar modes, could be used to create 3d structured light fields of chosen nano-scale characteristics which can be tailored for specific types of trapping objects or for innovative high resolution imaging.

References

1. Larocque H et al (2018) Reconstructing the topology of optical polarization knots. Nat Phys 14: 1079
2. Sugic D, Dennis MR (2018) Singular knot bundle in light. J Opt Soc Am A 35: 1987